# Fundamentals of Food Science Technology Processing and Preservation

# Fundamentals of Food Science Technology Processing and Preservation

N.K. Jain

CYBER TECH PUBLICATIONS
4264/3, Ansari Road, Daryaganj, New Delhi-110002 (India)
Ph.: 011-23244078, 011-43559448 Fax: 011-23280028
E-Mail: cyberpublicationsdelhi@yahoo.com
Website: www.cybertechpublications.com

# FUNDAMENTALS OF FOOD SCIENCE TECHNOLOGY PROCESSING AND PRESERVATION

**N.K. Jain**



First Edition 2011

Reprint 2024



*Published by :*

**G.S. Rawat** for **Cyber Tech Publications**
4264/3, Ansari Road, Daryaganj, New Delhi-110002 (India)
Ph.: 011-23244078, 011-43559448 Fax: 011-23280028
E-Mail: cyberpublicationsdelhi@yahoo.com
Website: www.cybertechpublications.com

*Printed at:* Asian Offset Printers, Delhi-53

# CONTENTS

# Preface

The cooking of food is a science as well as an art, and it depends for its success on known and established principles. In its full sense, *cookery* means not only the ability to follow a recipe, thereby producing a successfully cooked dish, but also the ability to select materials, a knowledge of the ways in which to prepare them, an understanding of their value for the persons for whom they are prepared, and ingenuity in serving foods attractively and in making the best use of food that may be left over from the previous meals, so that there will be practically no waste. Thus, while cookery in all its phases is a broad subject, it is one that truly belongs to woman, not only because of the pleasure she derives in preparing food for the members of her family, but because she is particularly qualified to carry on the work. While this refers especially to the preparation of food by subjecting edible materials to the action of heat, it involves much more.

The providing of food in the home is a matter that usually falls to the lot of the housewife; in fact, on her depends the wise use of the family income. This means, then, that whether a woman is earning her own livelihood and has only herself to provide for, or whether she is spending a part of some other person's income, as, for instance, her father's or her husband's, she should understand how to proportion her money so as to provide the essential needs, namely, food, clothing, and shelter. In considering the question of providing food, the housewife should set about to determine what three meals a day will cost, and in this matter she should be guided by the thought that the meals must be the best that can possibly be purchased for the amount of money allowed for food from the family income and that their cost must not exceed the allotment.

—*Author*

# 1

# Introduction

## THE PROBLEM OF FOOD

Without doubt, the greatest problem confronting the human race is that of food. In order to exist, every person must eat; but eating simply to keep life in the body is not enough. Aside from this, the body must be supplied with an ample amount of energy to carry on each day's work, as well as with the material needed for its growth, repair, and working power. To meet these requirements of the human body, there is nothing to take the place of *food,* not merely any kind, however, but the *right* kind. Indeed, so important is the right kind of food in the scheme of life that the child deprived of it neither grows nor increases in weight, and the adult who is unable to secure enough of it for adequate nourishment is deficient in nerve force and working power. If a person is to get the best out of life, the food taken into the body must possess real sustaining power and supply the tissues with the necessary building material; and this truth points out that there are facts and principles that must be known in order that the proper selection of food may be made, that it may be so prepared as to increase its value, and that economy in its selection, preparation, use, and care may be exercised.

Probably the most important of these principles is the *cooking of food.* While this refers especially to the preparation of food by subjecting edible materials to the action of heat, it involves much more. The cooking of food is a science as well as an art, and it depends for its success on known and established principles. In its full sense, *cookery* means not only the ability to follow a recipe, thereby producing a successfully cooked dish, but also the ability to select materials, a knowledge of the ways in which to prepare them, an understanding of their value for the persons for whom they are prepared, and ingenuity in serving foods attractively and in making the best use of food that may be left over from the previous meals, so that there will be practically no waste. Thus, while cookery in all its phases is a broad subject, it is one that truly belongs to woman, not only because of the pleasure she derives in preparing food for the members of her family, but because she is particularly qualified to carry on the work.

The providing of food in the home is a matter that usually falls to the lot of the housewife; in fact, on her depends the wise use of the family income. This means, then, that whether a woman is earning her own livelihood and has only herself to provide for, or whether she is spending a part of some other person's income, as, for instance, her father's or her husband's, she should understand how to proportion her money so as to provide the essential needs, namely, food, clothing, and shelter. In considering the question of providing food, the housewife should set about to determine what three meals a day will cost, and in this matter she should be guided by the thought that the meals must be the best that can possibly be purchased for the amount of money allowed for food from the family income and that their cost must not exceed the allotment. To a great extent she can control the cost of her foods by selecting them with care and then making good use of what her money has bought. It is only by constant thought and careful planning, however, that she will be able to keep within her means, and she will find that her greatest assistance lies in studying foods and the ways in which to prepare them.

A factor that should not be disregarded in the problem of food is *waste,* and so that the housewife can cope with it properly she should understand the distinction between waste and refuse. These terms are thought by some to mean the same thing and are often confused; but there is a decided difference between them. *Waste,* as applied to food, is something that could be used but is not, whereas *refuse* is something that is rejected because it is unfit for use. For example, the fat of meat, which is often eaten, is waste if it is thrown away, but potato parings, which are not suitable as food, are refuse.

In connection with the problem of waste, it may be well to know that leakage in the household is due to three causes. The first one is lack of knowledge on the part of the housekeeper as to the difference between waste and refuse and a consequent failure to market well. As an illustration, many housewives will reject turkey at a certain price a pound as being too expensive and, instead, will buy chicken at, say, 5 cents a pound less. In reality, chicken at 5 cents a pound less than the price of turkey is more expensive, because turkey, whose proportion of meat to bone is greater than that of chicken, furnishes more edible material; therefore, in buying chicken, they pay more for refuse in proportion to good material. The second cause for this leakage in the household is excessive waste in the preparation of food for the table, arising from the selection of the wrong cooking method or the lack of skill in cooking; and the third cause is the serving of too large quantities and a consequent waste of food left on individual plates and unfit for any other use in the home.

Another matter that constantly confronts the housewife is what foods she shall select for each day's meals. To be successful, all meals should be planned with the idea of making them wholesome and appetizing, giving them variety, and using the left-overs. Every woman should understand that food is cooked for both hygienic and esthetic reasons; that is, it must be made safe and wholesome for health's sake and must satisfy the appetite, which to a considerable degree is mental and, of course, is influenced by the appearance of the food. When the housewife knows how to cook ordinary foods well, she has an excellent foundation from which to obtain variety in the *diet*—by which in these lessons is meant the daily food and drink of any individual, and not something prescribed by a physician for a person who is ill—for then it is simply a matter of putting a little careful thought into the work she is doing in order to get ideas of new ways in which to prepare these same foods and of utilizing foodstuffs she has on hand. However, ample time must always be allowed for the preparation of meals, for no one can expect to produce tasty meals by rushing into the kitchen just before meal time and getting up the easiest thing in the quickest manner. Well-planned meals carefully prepared will stimulate interest in the next day's bill of fare and will prove extremely beneficial to all concerned.

In the practice of cookery it is also important that the meals be planned and the cooking done for the sake of building the human body and caring for it. As soon as any woman realizes that both the present and the future welfare of the persons for whom she is providing foods depend on so many things that are included in cookery, her interest in this branch of domestic science will increase; and in making a study of it she may rest assured that there is possibly no other calling that affords a more constant source of enjoyment and a better opportunity for acquiring knowledge, displaying skill, and helping others to be well and happy.

The fact that people constantly desire something new and different in the way of food offers the housewife a chance to develop her ingenuity along this line. Then, too, each season brings with it special foods for enjoyment and nourishment, and there is constant satisfaction in providing the family with some surprise in the form of a dish to which they are unaccustomed, or an old one prepared in a new or a better way. But the pleasure need not be one-sided, for the adding of some new touch to each meal will give as much delight to the one who prepares the food as to those who partake of it. When cookery is thought of in this way, it is really a creative art and has for its object something more than the making of a single dish or the planning of a single meal.

From what has been pointed out, it will readily be seen that a correct knowledge of cookery and all that it implies is of extreme importance to those who must prepare food for others; indeed, it is for just such persons—the housewife who must solve cookery problems from day to day, as well as girls and women who must prepare themselves to perform the duties with which they will be confronted when they take up the management of a household and its affairs—that these lessons in cookery are intended.

In the beginning of this course of study in cookery it is deemed advisable to call attention to the order in which the subject matter is presented. As will be seen before much progress is made, the lessons are arranged progressively; that is, the instruction begins with the essentials, or important fundamentals, of food—its selection, preparation, and care—and, from these as a foundation, advances step by step into the more complicated matters and minor details. The beginner eager to take up the actual work of cookery may feel that too much attention is given to preliminaries. However, these are extremely essential, for they are the groundwork on which the actual cooking of food depends; indeed, without a knowledge of them, very little concerning cookery in its various phases could be readily comprehended.

## THE PROBLEM OF FOOD

Each beginner in cookery is therefore urged to master every lesson in the order in which she receives it and to carry out diligently every detail. No lesson should be disregarded as soon as it is understood, for the instruction given in it bears a close relation to the entire subject and should be continually put into practice as progress is made. This thought applies with particular emphasis to the Sections relating to the essentials of cookery. These should be used in connection with all other Sections as books of reference and an aid in calling to mind points that must eventually become a part of a woman's cookery knowledge. By carrying on her studies systematically and following directions carefully, the beginner will find the cooking of foods a simple matter and will take delight in putting into practice the many things that she learns.

## SELECTION OF FOOD

### Matters Involved in Right Selection

Each one of the phases of cookery has its importance, but if success is to be achieved in this art, careful attention must be given to the selection of what is to be cooked, so as to determine its value and suitability. To insure the best selection, therefore, the housewife should decide whether the food material she purchases will fit the needs of the persons who are to eat it; whether the amount of labor involved in the preparation will be too great in proportion to the results obtained; whether the loss in preparation, that is, the proportion of refuse to edible matter, will be sufficient to affect the cost materially; what the approximate loss in cooking will be; whether the food will serve to the best advantage after it is cooked; and, finally, whether or not all who are to eat it will like it. The market price also is a factor that cannot be disregarded, for, as has been explained, it is important to keep within the limits of the amount that may be spent and at the same time provide the right kind of nourishment for each member of the family.

In order to select food material that will meet the requirements just set forth, three important matters must be considered; namely, the *substances* of which it is composed; its measure of energy-producing material, or what is called its *food, or fuel, value;* and its *digestion* and *absorption*. Until these are understood, the actual cost of any article of food cannot be properly determined, although its price at all times may be known.

However, before a study of any of these matters is entered into, it is necessary to know just what is meant by food and what food does for the body. As is well understood, the body requires material by which it may be built and its tissues repaired when they are torn down by work and exercise. In addition it requires a supply of heat to maintain it at normal temperature and provide it with sufficient energy to do the work required of it. The material that will accomplish these important things is food, which may therefore be regarded as anything that, when taken into the body, will build and repair its tissues or will furnish it with the energy required to do its work.

## FOOD SUBSTANCES

Although, as has just been stated, food may be considered as anything that the human engine can make over into tissue or use in living and working, not all foods are equally desirable any more than all materials are equally good in the construction of a steam engine and in the production of its working power. Those food substances which are the most wholesome and healthful are the ones to be chosen, but proper choice cannot be made unless the buyer knows of what the particular food consists and what it is expected to do. To aid in the selection of food, therefore, it is extremely necessary to become familiar with the five substances, constituents, or principles of which foods are made up; namely, water, mineral matter, or ash, protein, fat, and carbohydrate. A knowledge of these will help also in determining the cooking methods to adopt, for this depends on the effect that heat has on the various substances present in a food. Of course, so far as flavor is concerned, it is possible for the experienced cook to prepare many dishes successfully without knowing the effect of heat on the different food constituents; but to cook intelligently, with that success which makes for actual economy and digestibility, certain facts must be known concerning the food principles and the effect of dry and moist heat on foods.

### Water

Of the various constituents that are found in the human body, water occurs in the largest quantity. As a food substance, it is an extremely important feature of a person's diet. Its chief purpose is to replenish the liquids of the body and to assist in the digestion of food. Although nature provides considerable amounts of water in most foods, large quantities must be taken in the diet as a beverage. In fact, it is the need of the body for water that has led to the development of numerous beverages. Besides being necessary in building up the body and keeping it in a healthy condition, water has a special function to perform in cooking, as is explained later. Although this food substance is extremely essential to life, it is seldom considered in the selection of food, because, as has just been mentioned, nearly all foods contain water.

## Mineral Matter

Ranking next to water in the quantity contained in the human body is mineral matter. This constituent, which is also called *ash* or *mineral salts,* forms the main part of the body's framework, or skeleton. In the building and maintaining of the body, mineral salts serve three purposes—to give rigidity and permanence to the skeleton, to form an essential element of active tissue, and to provide the required alkalinity or acidity for the digestive juices and other secretions.

The origin and distribution of these mineral substances are of interest. Plants in their growth seize from the earth the salts of minerals and combine them with other substances that make up their living tissue. Then human beings, as well as other living creatures, get their supply of these needed salts from the plants that they take as food, this being the only form in which the salts can be thoroughly assimilated. These salts are not affected by cooking unless some process is used that removes such of them as are readily soluble in water. When this occurs, the result is usually waste, as, for instance, where no use is made of the water in which some vegetables are boiled. As is true of water, mineral matter, even though it is found in large quantities in the body, is usually disregarded when food is purchased. This is due to the fact that this important nutritive material appears in some form in nearly all foods and therefore does not necessitate the housewife's stopping to question its presence.

## Protein

The food substance known as protein is a very important factor in the growth and repair of the body; in fact, these processes cannot be carried on unless protein is present in the diet. However, while a certain quantity of protein is essential, the amount is not very large and more than is required is likely to be harmful, or, since the body can make no use of it, to be at least waste material. The principal sources of protein are lean meat, eggs, milk, certain grains, nuts, and the legumes, which include such foods as beans and peas. Because of the ease with which they are digested, meat, fish, eggs, and milk are more valuable sources of protein than bread, beans, and nuts. However, as the foods that are most valuable for proteins cost more than others, a mixed diet is necessary if only a limited amount of money with which to purchase foods is available.

So much is involved in the cooking of foods containing protein that the effect of heat on such foods should be thoroughly understood. The cooking of any food, as is generally understood, tends to break up the food and prepare it for digestion. However, foods have certain characteristics, such as their structure and texture, that influence their digestibility, and the method of cooking used or the degree to which the cooking is carried so affects these characteristics as to increase or decrease the digestibility of the food.

In the case of foods containing protein, unless the cooking is properly done, the application of heat is liable to make the protein indigestible, for the heat first coagulates this substance—that is, causes it to become thick—and then, as the heat increases, shrinks and hardens it. This fact is clearly demonstrated in the cooking of an egg, the white of which is the type of protein called *albumin*. In a raw egg, the albumin is nearly liquid, but as heat is applied, it gradually coagulates until it becomes solid. If the egg is cooked too fast or too long, it toughens and shrinks and becomes less palatable, less attractive, and less digestible. However, if the egg is properly cooked after the heat has coagulated the albumin, the white will remain tender and the yolk will be fine and mealy in texture, thus rendering it digestible.

Similar results, although not so evident to the sight, are brought about through the right or wrong way of cooking practically all other foods that contain much protein. Milk, whose principal ingredient is a protein known as *casein*, familiar as the curd of cheese, illustrates this fact very plainly. When it is used to make cottage cheese, heating it too long or to too high a degree will toughen the curd and actually spoil the texture of the product, which will be grainy and hard, instead of smooth and tender.

## Fats

The food substances just discussed—water, mineral matter, and protein—yield the materials required for building and repairing the tissues of the body, but as has been explained, the body also requires foods that produce energy, or working power. By far the greater part of the total solids of food taken into the body serve this purpose, and of these fats form a large percentage. Although fats make up such a large proportion of the daily food supply, they enter into the body composition to a less extent than do the food substances that have been explained. The fats commonly used for food are of both animal and vegetable origin, such as lard, suet, butter, cream, olive oil, nut oil, and cottonseed oil. The ordinary cooking temperatures have comparatively little effect on fat, except to melt it if it is solid. The higher temperatures decompose at least some of it, and thus liberate substances that may be irritating to the digestive tract.

## Carbohydrates

Like fats, the food substances included in the term carbohydrates supply the body with energy. However, fats and carbohydrates differ in the forms in which they supply energy, the former producing it in the most concentrated form and the latter in the most economical form.

So that the term *carbohydrate* may be clearly understood and firmly fixed in the mind, it is deemed advisable to discuss briefly the composition of the body and the food that enters it. Of course, in a lesson on cookery, not so much attention need be given to this matter as in a lesson on *dietetics*, which is a branch of hygiene that treats of diet; nevertheless, it is important that every person who prepares food for the table be familiar with the fact that the body, as well as food, is made up of a certain number of chemical elements, of which nitrogen, carbon, hydrogen, and oxygen form a large part. Protein owes its importance to the fact that of the various food substances it alone contains the element nitrogen, which is absolutely essential to the formation of any plant or animal tissue. The other three elements, carbon, hydrogen, and oxygen, go to make up the carbohydrates; in fact, it is from the names of these three elements that the term carbohydrate is derived.

The carbohydrates include the starches and sugars that are used and eaten in so many forms, and these contain the three elements mentioned, the hydrogen and oxygen contained in them being in the proportion that produces water. Thus, as will readily be seen, by separating the name into its parts—*carbo* (carbon) and *hydrate* (hydrogen and oxygen in the proportion of two parts of hydrogen and one of oxygen, that is, in the form of water)—carbohydrate is simply carbon united with water. While the facts just brought out have much to do with food economy, they are of interest here chiefly because they help to make clear the term carbohydrate, which, as will be admitted, is the only correct name for the food substance it represents. STARCH, one of the chief forms of carbohydrates, is found in only the vegetable kingdom. It is present in large quantities in the grains and in potatoes; in fact, nearly all vegetables contain large or small amounts of it. It is stored in the plant in the form of granules that lie within the plant cells. Cooking applied to starch changes it into a form that is digestible. Moist heat cooks the granules until they expand and burst and thus thicken the mass. Dry heat changes starch first into a soluble form and finally into what is called *dextrine*, this being the intermediate step in the changing of starch into sugar.

SUGAR, another important form of carbohydrate, is mainly of vegetable origin, except that which is found in milk and called *lactose*. This, together with the fat found in milk, supplies the child with energy before it is able to digest a variety of foods. The sap of various plants contains such large quantities of sugar that it can be crystalized out and secured in dry form. The liquid that remains is valuable as food, for, by boiling it down, it forms molasses. Sugar is also present in considerable amounts in all fruits, and much of it is in a form that can be assimilated, or taken up by the body, quickly. A sugar very similar to this natural fruit sugar is made from the starch of corn and is called *glucose*. Much of the carbohydrate found in vegetables, especially young, tender vegetables, is in the form of sugar, which, as the vegetables grow older, changes to starch.

Sugar melts upon the application of heat or, if it is in a melted condition, as sirup or molasses, it boils down and gives off water. When all the water has boiled away, the sugar begins to caramelize or become brown, and develops a characteristic flavor. If the cooking is continued too long, a dark-brown color and a bitter taste are developed. Because the sugar in fruits and vegetables is in solution, some of it is lost when they are boiled, unless, of course, the water in which they are cooked is utilized.

CELLULOSE is a form of carbohydrate closely related to starch. It helps to form the structure of plants and vegetables. Very little cellulose is digested, but it should not be ignored, because it gives the necessary bulk to the food in which it occurs and because strict attention must be paid to the cooking of it. As cellulose usually surrounds nutritive material of vegetable origin, it must be softened and loosened sufficiently by cooking to permit the nutritive material to be dissolved by the digestive juices. Then, too, in old vegetables, there is more starch and the cellulose is harder and tougher, just as an old tree is much harder than a sapling. This, then, accounts for the fact that rapid cooking is needed for some vegetables and slow cooking for others, the method and the time of cooking depending on the presence and the consistency of the cellulose that occurs in the food.

## IMPORTANCE OF A VARIETY OF FOODS

Every one of the five food substances just considered must be included in a person's diet; yet, with the exception of milk, no single food yields the right amounts of material necessary for tissue building and repair and for heat and energy. Even milk is in the right proportion, as far as its food substances are concerned, only for babies and very young children. It will thus be seen that to provide the body with the right foods, the diet must be such as to include all the food substances. In food selection, therefore, the characteristics of the various food substances must be considered well. Fats yield the most heat, but are the most slowly digested. Proteins and carbohydrates are more quickly digested than fats, but, in equal amounts, have less than half as much food value. Water and mineral salts do not yield heat, but are required to build tissue and to keep the body in a healthy condition. In addition, it is well to note that a well-balanced diet is one that contains all of the five food substances in just the right proportion in which the individual needs them to build up the body, repair it, and supply it with energy. What this proportion should be, however, cannot be stated offhand, because the quantity and kind of food substances necessarily vary with the size, age, and activity of each person.

## FOOD VALUE

Nearly all foods are complex substances, and they differ from one another in what is known as their *value,* which is measured by the work the food does in the body either as a tissue builder or as a producer of energy. However, in considering food value, the person who prepares food must not lose sight of the fact that the individual appetite must be appealed to by a sufficient variety of appetizing foods. There would be neither economy nor advantage in serving food that does not please those who are to eat it.

While all foods supply the body with energy, they differ very much in the quantity they yield. If certain ones were chosen solely for that purpose, it would be necessary for any ordinary person to consume a larger quantity of them than could be eaten at any one time. For instance, green vegetables furnish the body with a certain amount of energy, but they cannot be eaten to the exclusion of other things, because no person could eat in a day a sufficient amount of them to give the body all the energy it would need for that day's work. On the other hand, certain foods produce principally building material, and if they were taken for the purpose of yielding only energy, they would be much too expensive. Meats, for example, build up the body, but a person's diet would cost too much if meat alone were depended on to provide the body with all the energy it requires. Many foods, too, contain mineral salts, which, as has been pointed out, are needed for building tissue and keeping the body in a healthy condition.

To come to a correct appreciation of the value of different foods, it is necessary to understand the unit employed to measure the amount of work that foods do in the body. This unit is the CALORIE, or *calory,* and it is used to measure foods just as the inch, the yard, the pound, the pint, and the quart are the units used to measure materials and liquids; however, instead of measuring the food itself, it determines its food value, or fuel value. To illustrate what is meant, consider, for instance, 1/2 ounce of sugar and 1/2 ounce of butter. As far as the actual weight of these two foods is concerned, they are equal; but with regard to the work they do in the body they differ considerably. Their relative value in the body, however, can be determined if they are measured by some unit that can be applied to both. It is definitely known that both of them produce heat when they are oxidized, that is, when they are combined with oxygen; thus, the logical way of measuring them is to determine the quantity of heat that will be produced when they are eaten and united with oxygen, a process that causes the liberation of heat. The calorie is the unit by which this heat can be measured, it being the quantity of heat required to raise the temperature of 1 pint of water 4 degrees Fahrenheit, which is the name of the thermometer commonly used in the home. When burned as fuel, a square of butter weighing 1/2 ounce produces enough heat to raise 1 pint of water 400 degrees Fahrenheit, and it will yield the same amount of heat when it is eaten and goes through the slow process of oxidation in the body. On the other hand, 1/2 ounce of sugar upon being oxidized will produce only enough heat to raise the temperature of 1 pint of water about degrees Fahrenheit. Thus, as will be seen, 1/2 ounce of butter has a value of approximately 100 calories, whereas 1/2 ounce of sugar contains only about 57-1/2 calories.

Other foods yield heat in varying degrees, and their food value is determined in exactly the same way as that of butter and sugar. To give an idea of the composition of various food materials, as well as the number of calories that 1 pound of these food materials will yield, food charts published by the United States Department of Agriculture are here presented. As an understanding of these charts will prove extremely profitable in the selection of food, a careful study of them at this time is urged. In addition, reference to them should be made from time to time as the various kinds of foods are taken up, as the charts will then be more easily comprehended and their contents of more value.

## DIGESTION AND ABSORPTION OF FOOD

The third requirement in the selection of food, namely, its digestion and absorption, depends considerably on the persons who are to be fed. Food that is chosen for adults entirely would not be the same as that intended for both young persons and adults; neither would food that is to be fed to children or persons who are ill be the same as that which is to be served to robust adults who do a normal amount of work. No hard-and-fast rules can be laid down here for this phase of food selection, but as these lessons in cookery are taken up in turn, the necessary knowledge regarding digestibility will be acquired.

# 2

# Preparation of Food

## REASONS FOR COOKING FOOD

The term cookery, as has been explained, means the preparation of both hot and cold dishes for use as food, as well as the selection of the materials or substances that are to be cooked. The importance of cooking foods by subjecting them to the action of heat has been recognized for ages; and while it is true that there are many foods that appeal to the appetite in their raw state and still others that can be eaten either raw or cooked, there are several reasons why it is desirable to cook food, as will be seen from the following:

1. Cooking makes foods more palatable. This is true of such foods as meat, cereals, and many vegetables, which would be very unappetizing if they were eaten raw.
2. Cooking renders foods more digestible. For instance, the hard grains, such as wheat, and the dried vegetables, such as beans, cannot be readily digested unless they are softened by cooking. But while cooking makes such foods more digestible, it renders others more difficult of digestion, as in the case of eggs, the degree of digestibility depending somewhat on the cooking method used and the skill of the cook. An egg in an almost liquid form, or when only slightly cooked, as a soft-boiled egg, is more easily digested than when it becomes hardened by cooking. Then, too, a properly prepared hard-cooked egg is more digestible than an improperly cooked one, although the degree of hardness may be the same.
3. Cooking gives foods greater variety. The same food may be cooked by various methods and be given very different

tastes and appearances; on the other hand, it may be combined with a large number of other foods, so as to increase the variety of the dishes in which it is used. The large number of recipes found in cook books show the attempts that have been made to obtain variety in cooked dishes by the combining of different foods.

4. Cooking sterilizes foods either partly or completely. Many foods need partial or complete sterilization for safety. They must be completely sterlized if the germs that produce fermentation or putrefaction and thereby spoil food would be destroyed. This is done when fruits and vegetables are canned for keeping. Foods that are exposed to dust, flies, and improper handling should be thoroughly cooked in order to destroy any pathogenic germs that might be present. By such germs are meant disease-bearing germs. They differ from germs that produce fermentation and putrefaction, or spoiling, and that must in general be considered as a help, for these play an important part in the raising of bread and the preparation of various foods, as is pointed out later.
5. Cooking develops flavor in many foods. In the case of some vegetables, the flavoring substance is given off in the air by certain methods of cooking and a better flavor is thereby developed.

## METHODS OF COOKING

Food is cooked by the application of heat, which may be either moist or dry. While it is true that the art of cooking includes the preparation of material that is served or eaten raw, cooking itself is impossible without heat; indeed, the part of cooking that requires the most skill and experience is that in which heat is involved. Explicit directions for carrying on the various cooking processes depend on the kind of stove, the cooking utensils, and even the atmospheric conditions. In truth, the results of some processes depend so much on the state of the atmosphere that they are not successful on a day on which it is damp and heavy; also, as is well known, the stove acts perfectly on some days, whereas on other days it seems to have a stubborn will of its own. Besides the difficulties mentioned, the heat itself sometimes presents obstacles in the cooking of foods, to regulate it in such a way as to keep it uniform being often a hard matter. Thus, a dish may be spoiled by subjecting it to heat that is too intense, by cooking it too long, or by not cooking it rapidly enough. All these points must be learned, and the best way to master them is to put into constant practice the principles that are involved in cookery.

Without doubt, the first step in gaining a mastery of cookery is to become familiar with the different methods and processes, the ways in which they are applied, and the reasons for applying them. There are numerous ways of cooking food, but the principal processes are boiling, stewing, steaming, dry steaming, braizing, fricasseeing, roasting, baking, broiling, pan broiling, frying, and sautéing. Which one of these to use will depend on the food that is to be cooked and the result desired. If the wrong method is employed, there will be a waste of food material or the food will be rendered less desirable in flavor or tenderness. For example, it would be both wasteful and undesirable to roast a tough old fowl or to boil a tender young broiler.

The various methods of cookery just mentioned naturally divide themselves into three groups; namely, those involving dry heat, those requiring moist heat, and those in which hot fat is the cooking medium.

## Cooking With Dry Heat

Cooking with dry heat includes broiling, pan broiling, roasting, and baking; but, whichever of these processes is used, the principle is practically the same. In these processes the food is cooked by being exposed to the source of heat or by being placed in a closed oven and subjected to heated air. When dry heat is applied, the food to be cooked is heated to a much greater temperature than when moist heat is used.

## Broiling

The cooking process known as broiling consists in exposing directly to the source of heat the food that is to be cooked; that is, in cooking it over or before a clear bed of coals or a gas flame. The aim in broiling is to retain the juices of food and develop flavor. As it is a quick method, foods that are not tender, as, for example, tough meats, should not be broiled, because broiling does not help to render their fibers more tender. In applying this cooking process, which is particularly suitable for tender portions of meat and for young fowl, the food should be exposed to intense heat at first in order to sear all surfaces quickly and thus retain the juices. At the beginning of the cooking, the article that is being broiled should be turned often; then, as soon as the outside is browned, the heat should be reduced if possible, as with a gas stove, and the article allowed to cook until done. If the broiling is done over coals, it is necessary to continue the turning during the entire process. While broiling produces an especially good flavor in the foods to which it is applied, provided they are not tough, it is not the most economical way of cooking.

## Pan Broiling

Pan broiling is an adaptation of the broiling method. It consists in cooking food in a sissing-hot pan on top of the stove without the use of fat. In this process the surfaces of the steak, chop, or whatever the food may be, are quickly seared, after which the article is turned frequently and cooked more slowly until done. The object of pan broiling is the same as that of broiling, and it is resorted to, as a rule, when the fire is not in the right condition for broiling.

## Roasting

Originally, the term *to roast* meant to cook before a fire, because, before the time of stoves, practically all food was cooked in the fireplace. Food that was to be roasted was placed before the fire in a device that reflected heat, this device being open on the side toward the fire and closed on that toward the room. The roast was suspended in this device, slowly turned, and thus cooked by radiant heat—that is, heat given off in the form of direct rays—the principle being the same as that of broiling, but the application different. Nowadays, the term *roasting* is almost universally applied to the action of both hot air and radiant heat. However, much of what is called roasting is in reality baking. Foods cooked in the oven of an ordinary coal or gas range are really baked, although they are said to be roasted, and a covered roasting pan is a misnomer. Food must be exposed to the air in the process of cooking if it is to be roasted in the true sense.

It may be well to note that successful roasting or broiling depends more on the shape of the article to be roasted or broiled than on its weight. For this reason, thick, compact cuts of meat are usually selected for roasting and thin cuts for broiling. Good results also depend very much on the pan selected for the roasting process. One of the great aims in cooking should be to save or conserve all the food possible; that is, if by one process less waste in cooking results, it should be chosen rather than one that will result in loss at the end of the cooking process.

### Baking

By baking is meant cooking in a heated oven at temperatures ranging from 300 to 500 degrees Fahrenheit. As the term baking is frequently used in a wrong sense, the actual conditions of the process should be thoroughly understood. In both broiling and the original method of roasting, the heat is applied directly; that is, the food is exposed directly to the source of heat. Actual baking differs from these processes in that it is done in a closed oven or by means of heated air. Starchy foods, such as bread, cakes, and pastry, are nearly always baked, and gradually other foods, such as meats, fish, and vegetables are being subjected to this method of cooking. In fact, persons who are skilled in cooking use the oven more and more for things that they formerly thought had to be cooked in other ways. But the name that is applied to the process depends somewhat on custom, for while meat that is cooked in the oven is really baked, it is usually termed roasted meat. It seems strange, but it is nevertheless true, that ham cooked in the oven has always been termed baked, while turkey cooked in exactly the same way is said to be roasted.

## COOKING WITH MOIST HEAT

The methods of cooking with moist heat, that is, through the medium of water, are boiling, simmering, steaming, dry steaming, and braizing. In every one of these processes, the effect of moist heat on food is entirely different from that of dry heat. However, the method to be selected depends to a great extent on the amount of water that the food contains. To some foods much water must be added in the cooking process; to others, only a little or none at all. If food is not placed directly in large or small quantities of water, it is cooked by contact with steam or in a utensil that is heated by being placed in another containing boiling water, as, for example, a double boiler.

As water is such an important factor in cooking with moist heat, something concerning its nature and use should be understood. Therefore, before considering the moist-heat cooking processes in detail, the function of water in the body and in cooking and also the kinds of water are discussed.

## Function of Water in the Body

Water supplies no energy to the body, but it plays a very important part in nutrition. In fact, its particular function in the body is to act as a solvent and a carrier of nutritive material and waste. In doing this work, it keeps the liquids of the body properly diluted, increases the flow of the digestive juices, and helps to carry off waste material. However, its ability to perform these necessary functions in the right way depends on its quality and its safety.

## Kinds of Water

Water is either hard or soft. As it falls from the clouds, it is pure and soft until it comes in contact with gases and solids, which are dissolved by it and change its character. It is definitely known that the last of the water that falls in a shower is much better than the first, as the first cleanses not only the air, but the roofs and other things with which it comes in contact. In passing through certain kinds of soil or over rocks, water dissolves some of the minerals that are contained there and is thus changed from soft to hard water. If sewage drains into a well or water supply, the water is liable to contain bacteria, which will render it unfit and unsafe for drinking until it is sterilized by boiling. Besides rain water and distilled water, there is none that is entirely soft; all other waters hold certain salts in solution to a greater or less degree.

The quality of hardness, which is present in nearly all water, is either temporary or permanent. Water is temporarily hard when it contains soluble lime, which is precipitated, that is, separated from it, upon boiling. Every housewife who uses a teakettle is familiar with this condition. The lime precipitated day after day clings to the sides of the vessel in which the water is boiled, and in time they become very thickly coated. Permanent hardness is caused by other compounds of lime that are not precipitated by boiling the water. The only way in which to soften such water is to add to it an alkali, such as borax, washing soda, or bicarbonate of soda.

## Uses of Water in Cooking

It is the solvent, or dissolving, power of water that makes this liquid valuable in cooking, but of the two kinds, soft water is preferable to hard, because it possesses greater solvent power. This is due to the fact that hard water has already dissolved a certain amount of material and will therefore dissolve less of the food substances and flavors when it is used for cooking purposes than soft water, which has dissolved nothing. It is known, too, that the flavor of such beverages as tea and coffee is often greatly impaired by the use of hard water. Dried beans and peas, cereals, and tough cuts of meat will not cook tender so readily in hard water as in soft, but the addition of a small amount of soda during the cooking of these foods will assist in softening them.

Water is used in cooking chiefly for extracting flavors, as in the making of coffee, tea, and soups; as a medium for carrying flavors and foods in such beverages as lemonade and cocoa; for softening both vegetable and animal fiber; and for cooking starch and dissolving sugar, salt, gelatine, etc. In accomplishing much of this work, water acts as a medium for conveying heat.

## Boiling

As applied to cooking, boiling means cooking foods in boiling water. Water boils when its temperature is raised by heat to what is commonly termed its *boiling point.* This varies with the atmospheric pressure, but at sea level, under ordinary conditions, it is always 212 degrees Fahrenheit. When the atmospheric pressure on the surface of the water is lessened, boiling takes place at a lower temperature than that mentioned, and in extremely high altitudes the boiling point is so lowered that to cook certain foods by means of boiling water is difficult. As the water heats in the process of boiling, tiny bubbles appear on the bottom of the vessel in which it is contained and rise to the surface. Then, gradually, the bubbles increase in size until large ones form, rise rapidly, and break, thus producing constant agitation of the water.

Boiling has various effects on foods. It toughens the albumin in eggs, toughens the fiber and dissolves the connective tissues in meat, softens the cellulose in cereals, vegetables, and fruits, and dissolves other substances in many foods. A good point to bear in mind in preparing foods by boiling is that slowly boiling water has the same temperature as rapidly boiling water and is therefore able to do exactly the same work. Keeping the gas burning full heat or running the fire hard to keep the water boiling rapidly is therefore unnecessary; besides, it wastes fuel without doing the work any faster and sometimes not so well. However, there are several factors that influence the rapidity with which water may be brought to the boiling point; namely, the kind of utensil used, the amount of surface exposed, and the quantity of heat applied. A cover placed on a saucepan or a kettle in which food is to be boiled retains the heat, and thus causes the temperature to rise more quickly; besides, a cover so used prevents a loss of water by condensing the steam as it rises against the cover. As water boils, some of it constantly passes off in the form of steam, and for this reason sirups or sauces become thicker the longer they are cooked. The evaporation takes place all over the surface of the water; consequently, the greater the surface exposed, the more quickly is the quantity of water decreased during boiling. Another point to observe in the boiling process is that foods boiled rapidly in water have a tendency to lose their shape and are reduced to small pieces if allowed to boil long enough.

Besides serving to cook foods, boiling also renders water safe, as it destroys any germs that may be present. This explains why water must sometimes be boiled to make it safe for drinking. Boiled water, as is known, loses its good taste. However, as this change is brought about by the loss of air during boiling, the flavor can be restored and air again introduced if the water is shaken in a partly filled jar or bottle, or beaten vigorously for a short time with an egg beater.

## Simmering or Stewing

The cooking process known as simmering, or stewing, is a modification of boiling. By this method, food is cooked in water at a temperature below the boiling point, or anywhere from 185 to 200 degrees Fahrenheit. Water at the simmering point always moves gently—never rapidly as it does in boiling. Less heat and consequently less fuel are required to cook foods in this way, unless, of course, the time consumed in cooking the food at a low temperature is much greater than that consumed in cooking it more rapidly.

Aside from permitting economy in the use of fuel, simmering, or stewing, cooks deliciously certain foods that could not be selected for the more rapid methods. For example, tough cuts of meat and old fowl can be made tender and tasty by long cooking at a low temperature, for this method tends to soften the fiber and to develop an excellent flavor. Tough vegetables, too, can be cooked tender by the simmering process without using so much fuel as would be used if they were boiled, for whatever method is used they require long cooking. Beets, turnips, and other winter vegetables should be stewed rather than boiled, as it is somewhat difficult to cook them tender, especially in the late winter and early spring. If dry beans and peas are brought to the simmering point and then allowed to cook, they can be prepared for the table in practically the same length of time and without so much fuel as if they boiled continuously.

## Steaming

As its name implies, steaming is the cooking of food by the application of steam. In this cooking process, the food is put into a *steamer,* which is a cooking utensil that consists of a vessel with a perforated bottom placed over one containing water. As the water boils, steam rises and cooks the food in the upper, or perforated, vessel. Steamers are sometimes arranged with a number of perforated vessels, one on top of the other. Such a steamer permits of the cooking of several foods at the same time without the need of additional fuel, because a different food may be placed in each vessel.

Steaming is preferable to boiling in some cases, because by it there is no loss of mineral salts nor food substances; besides, the flavor is not so likely to be lost as when food is boiled. Vegetables prepared in this way prove very palatable, and very often variety is added to the diet by steaming bread, cake, and pudding mixtures and then, provided a crisp outside is desired, placing them in a hot oven to dry out the moist surface.

## Dry Steaming

Cooking foods in a vessel that is suspended in another one containing boiling water constitutes the cooking method known as dry steaming. The double boiler is a cooking utensil devised especially for carrying on this process. The food placed in the suspended, or inner, vessel does not reach the boiling point, but is cooked by the transfer of heat from the water in the outside, or lower, vessel. A decided advantage of this method is that no watching is required except to see that the water in the lower vessel does not boil away completely, for as long as there is water between the food and the fire, the food will neither boil nor burn.

Because of the nature of certain foods, cooking them by this process is especially desirable. The flavor and consistency of cereals and foods containing starch are greatly improved by long cooking in this way. Likewise, custards and mixtures containing eggs can be conveniently cooked in a double boiler, because they do not require a high temperature; in fact, their texture is spoiled if they are cooked at the boiling point. To heat milk directly over the flame without scorching it is a difficult matter, and, on the other hand, boiled milk is hard to digest. Because of these facts, food containing milk should not be boiled, but should be cooked at a lower temperature in a double boiler.

### Braizing

Cooking meat in an oven in a closed pan with a small quantity of water constitutes braizing. This cooking process might be called a combination of stewing and baking, but when it is properly carried out, the meat is placed on a rack so as to be raised above the water, in which may be placed sliced vegetables. In this process the meat actually cooks in the flavored steam that surrounds it in the hot pan. The so-called double roasting pans are in fact braizing pans when they are properly used. A pot roast is the result of a modification of the braizing method.

## COOKING WITH HOT FAT

Of the three mediums of conveying heat to food, namely, hot air, hot water, and hot fat, that of hot fat renders food the least digestible. Much of this difficulty, however, can be overcome if an effort is made to secure as little absorption of the fat as possible. If the ingredients of the food are properly mixed before applying the fat and if the fat is at the right temperature, good results can be obtained by the various methods of cooking with hot fat, which are frying, sautéing, and fricasseeing.

### Frying

By frying is meant the cooking of food in deep fat at a temperature of 350 to 400 degrees Fahrenheit. Any kind of fat that will not impart flavor to the food may be used for frying, but the vegetable oils, such as cottonseed oils, combinations of coconut and cottonseed oils, and nut oils, are preferable to lards and other animal fats, because they do not burn so easily. Foods cooked in deep fat will not absorb the fat nor become greasy if they are properly prepared, quickly fried, and well drained on paper that will absorb any extra fat.

### Sautéing

Browning food first on one side and then on the other in a small quantity of fat is termed sautéing. In this cooking process, the fat is placed in a shallow pan, and when it is sufficiently hot, the food is put into it. Foods that are to be sautéd are usually sliced thin or cut into small pieces, and they are turned frequently during the process of cooking. All foods prepared in this way are difficult to digest, because they become more or less hard and soaked with fat. Chops and thin cuts of meat, which are intended to be pan-broiled, are really sautéd if they are allowed to cook in the fat that fries out of them.

### Fricasseeing

A combination of sautéing and stewing results in the cooking process known as fricasseeing. This process is used in preparing such foods as chicken, veal, or game, but it is more frequently employed for cooking fowl, which, in cookery, is the term used to distinguish the old of domestic fowls from chickens or pullets. In fricasseeing, the meat to be cooked is cut into pieces and sautéd either before or after stewing; then it is served with a white or a brown sauce. Ordinarily, the meat should be browned first, unless it is very tough, in order to retain the juices and improve the flavor. However, very old fowl or tough meat should be stewed first and then browned.

## HEAT FOR COOKING

Inasmuch as heat is so important a factor in the cooking of foods, it is absolutely necessary that the person who is to prepare them be thoroughly familiar with the ways in which this heat is produced. The production of heat for cooking involves the use of fuels and stoves in which to burn them, as well as electricity, which serves the purpose of a fuel, and apparatus for using electricity. In order, therefore, that the best results may be obtained in cookery, these subjects are here taken up in detail.

Probably the first fuel to be used in the production of heat for cooking was wood, but later such fuels as peat, coal, charcoal, coke, and kerosene came into use. Of these fuels, coal, gas, and kerosene are used to the greatest extent in the United States. Wood, of course, is used considerably for kindling fires, and it serves as fuel in localities where it is abundant or less difficult to procure than other fuels. However, it is fast becoming too scarce and too expensive to burn. If it must be burned for cooking purposes, those who use it should remember that dry, hard wood gives off heat at a more even rate than soft wood, which is usually selected for kindling. Electricity is coming into favor for supplying heat for cooking, but only when it can be sold as cheaply as gas will its use in the home become general.

The selection of a stove to be used for cooking depends on the fuel that is to be used, and the fuel, in turn, depends on the locality in which a person lives. However, as the fuel that is the most convenient and easily obtained is usually the cheapest, it is the one to be selected, for the cost of the cooked dish may be greatly increased by the use of fuel that is too expensive. In cooking, every fuel should be made to do its maximum amount of work, because waste of fuel also adds materially to the cost of cooking and, besides, it often causes great inconvenience. For example, cooking on a red-hot stove with a fire that, instead of being held in the oven and the lids, overheats the kitchen and burns out the stove not only wastes fuel and material, but also taxes the temper of the person who is doing the work. From what has just been said, it will readily be seen that a knowledge of fuels and apparatus for producing heat will assist materially in the economical production of food, provided, of course, it is applied to the best advantage.

## COAL AND COKE

### Varieties of Coal

Possibly the most common fuel used for cooking is coal. This fuel comes in two varieties, namely, *anthracite*, or *hard coal*, and *bituminous*, or *soft coal*. Their relative cost depends on the locality, the kind of stove, and an intelligent use of both stove and fuel. Hard coal costs much more in some places than soft coal, but it burns more slowly and evenly and gives off very little smoke. Soft coal heats more rapidly than hard coal, but it produces considerable smoke and makes a fire that does not last so long. Unless a stove is especially constructed for soft coal, it should not be used for this purpose, because the burning of soft coal will wear it out in a short time. The best plan is to use each variety of coal in a stove especially constructed for it, but if a housewife finds that she must at times do otherwise, she should realize that a different method of management and care of the stove is demanded.

## Sizes of Coal

As the effect of coal on the stove must be taken into consideration in the buying of fuel, so the different sizes of hard coal must be known before the right kind can be selected. The sizes known as *stove* and *egg coal,* which range from about 1-3/8 to 2-3/4 inches in diameter, are intended for a furnace and should not be used in the kitchen stove for cooking purposes. Some persons who know how to use the size of coal known as *pea,* which is about 1/2 to 3/4 inch in diameter, like that kind, whereas others prefer the size called *chestnut,* which is about 3/4 inch to 1-3/8 inches in diameter. In reality, a mixture of these two, if properly used, makes the best and most easily regulated kitchen coal fire.

## Quality of Coal

In addition to knowing the names, prices, and uses of the different kinds of coal, the housewife should be able to distinguish poor coal from good coal. In fact, proper care should be exercised in all purchasing, for the person who understands the quality of the thing to be purchased will be more likely to get full value for the money paid than the one who does not. About coal, it should be understood that good hard coal has a glossy black color and a bright surface, whereas poor coal contains slaty pieces. The quality of coal can also be determined from the ash that remains after it is burned. Large chunks or great quantities of ash indicate a poor quality of coal, and fine, powdery ash a good quality. Of course, even if the coal is of the right kind, poor results are often brought about by the bad management of a fire, whether in a furnace or a stove. Large manufacturing companies, whose business depends considerably on the proper kind of fuel, buy coal by the heat units—that is, according to the quantity of heat it will give off—and at some future time this plan may have to be followed in the private home, unless some other fuel is provided in the meantime.

Mixed with poor coal are certain unburnable materials that melt and stick together as it burns and form what are known as *clinkers*. Clinkers are very troublesome because they often adhere to the stove grate or the lining of the firebox. They generally form during the burning of an extremely hot fire, but the usual temperature of a kitchen fire does not produce clinkers unless the coal is of a very poor quality. Mixing oyster shells with coal of this kind often helps to prevent their formation.

### Coke

Another fuel that is sometimes used for cooking is coke. Formerly, coke was a by-product in the manufacture of illuminating gas, but now it is manufactured from coal for use as a fuel. Because of the nature of its composition, coke produces a very hot fire and is therefore favorable for rapid cooking, such as broiling. However, it is used more extensively in hotels and institutions than in kitchens where cooking is done on a small scale.

## GAS

### Value of Gas as Fuel

As a fuel for cooking purposes, gas, both *artificial* and *natural*, is very effective, and in localities where the piping of gas into homes is possible it is used extensively. Of the two kinds, artificial gas produces the least heat; also, it is the most expensive, usually costing two or three times as much as natural gas. Both are very cheap, however, considering their convenience as a kitchen fuel. Heat from gas is obtained by merely turning it on and igniting it, as with a lighted match. Its consumption can be stopped at once by closing off the supply, or it can be regulated as desired and in this way made to give the exact amount of heat required for the method of cookery adopted. Neither smoke nor soot is produced in burning gas if the burners of the gas stove are adjusted to admit the right amount of air, and no ashes nor refuse remain to be disposed of after gas has been burned. Because gas is so easily handled, good results can be obtained by those who have had very little experience in using it, and with study and practice results become uniform and gas proves to be an economical fuel.

### Measurement of Gas

Gas is measured by the cubic foot, and a definite price is

charged for each 1,000 cubic feet. To determine the quantity used, it is passed through what is called a meter, which measures as the gas burns. It is important that each housewife be able to read the amount registered by the meter, so that she can compare her gas bill with the meter reading and thus determine whether the charges are correct. If only the usual amount of gas has been consumed and the bill does not seem to be correct or is much larger than it has been previously, the matter should be reported to the proper authorities, for the meter may be out of order and in need of repair.

## Reading a Gas Meter

To register the quantity of gas that is consumed, a gas meter, as is shown in Fig. 1, is provided with three large dials, each of which has ten spaces over which the hand, or indicator, passes to indicate the amount of gas consumed, and with one small dial, around which the hand makes one revolution every time 2 cubic feet of gas is consumed. This small dial serves to tell whether gas is leaking when the stoves and lights are not turned on. Above each large dial is an arrow that points out the direction in which to read, the two outside ones reading toward the right and the center one toward the left; also, above each dial is lettered the quantity of gas that each dial registers, that at the right registering 1,000 cubic feet, that in the center 10,000 cubic feet, and that at the left 100,000 cubic feet. To read the dials, begin at the left, or the 100,000 dial, and read toward the right. In each instance, read the number over which the hand has passed last. For instance, when, as in Fig. 1, the hand lies between 5 and 6 on the left dial, 5 is read; on the center dial, when the hand lies between 5 and 6, 5 is read also; and on the right dial, when the hand lies between 2 and 3, the 2, which is really 200, is read.

To compute the quantity of gas used, the dials are read from left to right and the three readings are added.

Then, in order to determine the quantity burned since the previous reading, the amount registered at that time, which is always stated on the gas bill, must be subtracted from the new reading.

To illustrate the manner in which the cost of gas consumed may be determined, assume that gas costs 90 cents per 1,000 cubic feet, that the previous reading of the gas meter, say on May 15, was 52,600 cubic feet, and that on June 15 the meter registered as shown in Fig. 1. As was just explained, the left dial of the meter reads 5, the center dial 5, and the right dial 200. Therefore, put these figures down so that they follow one another, as 5-5-200. This means then that the reading on June 15 is 55,200 cubic feet. With this amount ascertained, subtract from it the previous reading, or 52,600 cubic feet, which will give 2,600 cubic feet, or the quantity of gas burned from May 15 to June 15. Since gas costs 90 cents per 1,000 cubic feet, the cost of the amount burned, or 2,600 cubic feet, may be estimated by dividing 2,600 cubic feet by 1,000 and multiplying the result by 90; thus 2,600 ÷ 1,000 = 2.6, and 2.6 x.90 = 2.34

## Prepayment Meters

In many places, gas concerns install what are called prepayment meters; that is, meters in which the money is deposited before the gas is burned. Such meters register the consumption of the gas in the same way as the meters just mentioned, but they contain a receptacle for money. A coin, generally a quarter, is dropped into a slot leading to this receptacle, and the amount of gas sold for this sum is then permitted to pass through as it is needed. When this amount of gas has been burned, another coin must be inserted in the meter before more gas will be liberated.

## KEROSENE

In communities where gas is not available, kerosene, which is produced by the refinement of petroleum, is used extensively as a fuel for cooking, especially in hot weather when the use of a coal or a wood stove adds materially to the discomfort of the person who does the cooking.

Kerosene is burned in stoves especially designed for its use, and while it is a cheap fuel it is not always the same in quality. It contains water at all times, but sometimes the proportion of water is greater than at others. The greater the amount of water, the less fuel will be contained in each gallon of kerosene. The quality of kerosene can be determined by checking up the length of time the stove will burn on a specified quantity of each new purchase of it.

Another product of the refinement of petroleum is *gasoline*. However, it is not used so extensively for fuel as kerosene, because it is more dangerous and more expensive.

## ELECTRICITY

The use of electricity for supplying heat for cooking is very popular in some homes, especially those which are properly wired, because of its convenience and cleanliness and the fact that the heat it produces can be applied direct. The first electrical cooking apparatus was introduced at the time of the World's Fair in

Chicago, in 1892, and since that time rapid advancement has been made in the production of suitable apparatus for cooking electrically. Electricity would undoubtedly be in more general use today if it were possible to store it in the same way as artificial gas, but as yet no such method has been devised and its cost is therefore greater. Electricity is generated in large power plants, and as it is consumed in the home for lighting and cooking it passes through a meter, which indicates the quantity used in much the same manner as a gas meter. It will be well, therefore, to understand the way in which an electric meter is read, so that the bills for electricity can be checked.

### Reading an Electric Meter

An electric meter, which is similar in appearance to a gas meter, consists of three or four dials, which are placed side by side or in the shape of an arc. In the usual type, and which consists of four dials placed side by side, each one of the dials contains ten spaces and a hand, or indicator, that passes over numbers ranging from to 9 to show the amount of electricity used. The numbers on the dials represent *kilowatt-hours*, a term meaning the energy resulting from the activity of 1 kilowatt for 1 hour, or 1 watt, which is the practical unit of electrical power, for 1,000 hours. Since 1,000 hours equal 1 kilowatt, 1,000 watt-hours equal 1 kilowatt-hour. It will be observed from the accompanying illustration that the dial on the extreme right has the figures reading in a clockwise direction, that is, from right to left, the second one in a counter-clockwise direction, or from left to right, the third one in a clockwise direction, and the fourth one in a counter-clockwise direction; also that above each dial is indicated in figures the number of kilowatt-hours that one complete revolution of the hand of that dial registers.

To read the meter, begin at the right-hand dial and continue to the left until all the dials are read and set the numbers down just as they are read; that is, from right to left. In case the indicator does not point directly to a number, but is somewhere between two numbers, read the number that it is leaving. In the second dial, the hand lies between and 1, and as it is leaving 0, this number should be read and placed to the left of the first one read, which gives 04. The hand on the third dial points exactly to 6; so 6 should be read for this dial and placed directly before the numbers read for the first and second dials, thus, 604. On the fourth and last dial, the indicator is between 4 and 5; therefore 4, which is the number it is leaving should be read and used as the first figure in the entire reading, which is 4,604.

After the reading of the electric meter has been ascertained, it is a simple matter to determine the electricity consumed since the last reading and the amount of the bill. For instance, assume that a meter registers the number of kilowatt-hours, or 4,604, and that at the previous reading it registered 4,559. Merely subtract the previous reading from the last one, which will give 45, or the number of kilowatt-hours from which the bill for electricity is computed. If electricity costs 3 cents a kilowatt-hour, which is the price charged in some localities, the bill should come to 45 X.03 or $1.35.

## PRINCIPLE OF STOVES

Before stoves for cooking came into use in the home, food was cooked in open fireplaces. Even when wood was the only fuel known, a stove for burning it, called the Franklin stove, was invented by Benjamin Franklin, but not until coal came into use as fuel were iron stoves made. For a long time stoves were used mainly for heating purposes, as many housewives preferred to cook at the open fireplace. However, this method of cooking has practically disappeared and a stove of some kind is in use for cooking in every home.

For each fuel in common use there are many specially constructed stoves, each having some advantageous feature; yet all stoves constructed for the same fuel are practically the same in principle. In order that fuel will burn and produce heat, it must have air, because fuel, whether it is wood, coal, or gas, is composed largely of *carbon* and air largely of *oxygen,* and it is the rapid union of these two chemical elements that produces heat. Therefore, in order that each stove may work properly, some way in which to furnish air for the fire in the firebox must be provided. For this reason, every stove for cooking contains passageways for air and is connected with a chimney, which contains a flue, or passage, that leads to the outer air.

When the air in a stove becomes heated, it rises, and as it ascends cold air rushes through the passageways of the stove to take its place. It is the flue, however, that permits of the necessary draft and carries off unburned gases. At times it is necessary to regulate the amount of air that enters, and in order that this may be done each stove is provided with *dampers*. These devices are located in the air passages and they are so designed as to close off the air or allow the desired amount to enter. By means of these dampers it is possible also to force the heat around the stove oven, against the top of the stove, or up the chimney flue. A knowledge of the ways in which to manipulate these dampers is absolutely necessary if correct results are to be obtained from a stove. The flue, however, should receive due consideration. If a stove is to give its best service, the flue, in addition to being well constructed, should be free from obstructions and kept in good condition. Indeed, the stove is often blamed for doing unsatisfactory work when the fault is really with the flue.

Probably one of the most important things considered in the construction of stoves is the economizing of fuel, for ever since the days of the fireplace there has been more or less of a tendency to save fuel for cooking, and as the various kinds grow scarcer, and consequently more expensive, the economical use of fuel becomes a necessity. While most stoves for cooking purposes are so constructed as to save fuel, many of them do not, especially if the method of caring for them is not understood. Any housewife, however, can economize in the use of fuel if she will learn how the stove she has must be operated; and this can be done by following closely the directions that come with the stove when it is purchased. Such directions are the best to follow, because they have been worked out by the manufacturer, who understands the right way in which his product should be operated.

## COAL, STOVES AND THEIR OPERATION

### General Construction

The principal parts of such a stove, which is commonly referred to as a *cook stove*, or range, are the firebox *a*; the grate *b*;

the ash pit *c*, which usually contains an ash-pan *d;* the oven *e;* the dampers *f, g, h,* and *i;* the flue opening *j* and flue *k;* openings in the top and suitable lids, not shown, for kettles and pans; and the air space extending from the firebox around three sides of the oven, as shown by the arrows. To prevent the stove from wearing out rapidly, the firebox, in which the fuel is burned, is lined with a material, such as fireclay, that will withstand great heat. The fire in the firebox is supported by the grate, which is in the form of metal teeth or bars, so as to permit air to pass through the fuel from underneath. The grate is usually so constructed that when the fire is raked it permits burnt coal or ashes to fall into the ash-pan, by means of which they can be readily removed from the stove. The oven, which lies directly back of the firebox and is really an enclosed chamber in which food may be cooked, receives its heat from the hot air that passes around it. The dampers are devices that control the flow of air in and out of the stove. Those shown at *f* and *g* serve to admit fresh air into the stove or to keep it out, and those shown at *h* and *i* serve to keep heated air in the stove or to permit it to pass out through the flue.

## Building a Coal Fire

To build a coal fire is a simple matter. So that the draft will be right for rapid combustion, it is first necessary to close the dampers *f* and *h* and to open the bottom damper *g* and the chimney damper *i*. With these dampers arranged, place crushed paper or shavings on the grate; then on top of the paper or shavings place kindling, and on top of the kindling put a small quantity of coal. Be careful to place the fuel on the grate loosely enough to permit currents of air to pass through it, because it will not burn readily if it is closely packed.

Light the fire by inserting a flame from below. When this is done, the flame will rise and ignite the kindling, and this, in turn, will cause the coal to take fire. When the fire is burning well, close the dampers *g* and *i* so that the fuel will not burn too rapidly and the heat will surround the oven instead of passing up the chimney; also, before too much of the first supply of coal is burned out, add a new supply, but be sure that the coal is sufficiently ignited before the new supply is added so as not to smother the fire. If only a thin layer is added each time, this danger will be removed. Experience has proved that the best results are secured if the fire is built only 4 inches high. When hot coals come near the top of the stove, the lids are likely to warp and crack from the heat and the cooking will not be done any more effectively. Another thing to avoid in connection with a fire is the accumulation of ashes. The ash-pan should be kept as nearly empty as possible, for a full ash-pan will check the draft and cause the grate in the firebox to burn out.

## Adjusting the Dampers

To get the best results from a cook stove, and at the same time overcome the wasting of fuel, the ways in which to adjust the dampers should be fully known. If it is desired to heat the oven for baking, close dampers *f* and *i* and open dampers *g* and *h*. With the dampers so arranged, the heated air above the fire is forced around the oven and up the flue. A study of this diagram will readily show that the lower left-hand corner of the oven is its coolest part, since the heated air does not reach this place directly, and that the top center is the hottest part, because the hottest air passes directly over this portion of the oven and the heated air in the oven rises to it.

If it is desired to heat the surface of the stove, so that cooking may be done on top of it, close dampers *f*, *h*, and *i* and open damper *g*. With the dampers so arranged, the heated air does not pass around the oven, but is confined in the space above it and the firebox, as shown in Fig. 4. While the damper *i* in the flue is closed in order to confine the heated air as much as possible to the space under the top of the stove, it contains openings that allow just enough air to pass up the flue to maintain the draft necessary for combustion. When the dampers are arranged as mentioned, the hottest place on the surface of the stove is between the firebox and the stovepipe, and the coolest place is behind the damper *h*.

## Banking a Coal Fire

To economize in the use of fuel, as well as to save the labor involved in building a new fire, it is advisable to keep a fire burning low from one meal to another and from one day to the next. As the nature of hard coal is such that it will hold fire for a long time, this can be done by what is called *banking* the fire. To achieve this, after the fire has served to cook a meal, shake the ashes out of the grate so that the glowing coals are left. Then put fresh coal on this bed of coals, and, with the dampers arranged as for building a new fire, allow the coal to burn well for a short time. Finally, cover the fire with a layer of fine coal and adjust the dampers properly; that is, close dampers *g* and *h* and open dampers *f* and *i*. If the banking is carefully done the fire should last 8 or 10 hours without further attention. Care should be taken, however, to use sufficient coal in banking the fire, so that when it is to be used again the coal will not be completely burned, but enough burning coals will remain to ignite a fresh supply. When the fire is to be used again, rake it slightly, put a thin layer of coal over the top, and arrange the dampers as for starting a fire. As soon as this layer of coal has begun to burn, add more until the fire is in good condition.

## GAS STOVES AND THEIR OPERATION

### Gas Ranges

A gas stove for cooking, or *gas range*, as it is frequently called, consists of an oven, a broiler, and several burners over which are plates to hold pans, pots, and kettles in which food is to be cooked. As is true of a coal range, a gas range also requires a flue to carry off the products of unburned gas. Gas stoves, or ranges, are of many makes, but in principle all of them are practically the same; in fact, the chief difference lies in the location or arrangement of the oven, broiler, and burners. The oven *a* of this stove is located above the top of the stove, instead of below it, as in some stoves. An oven so located is of advantage in that it saves stooping or bending over. The door of this oven contains a glass, which makes it possible to observe the food baking inside without opening the door and thereby losing heat. The broiler *b*, which may also be used as a toaster, is located directly beneath the oven, and to the right are the burners *c* for cooking. The gas for these parts is contained in the pipe *d*, which is connected to a pipe joined to the gas main in the street. To get heat for cooking it is simply necessary to turn on the stop-cocks and light the gas. The four burners are controlled by the stop-cocks *e*, and the oven and the broiler by the stop-cock *f*. The stove is also equipped with a simmering burner for the slow methods of cooking on top of the stove, gas to this burner being controlled by the stop-cock *g*. To catch anything that may be spilled in cooking, there is a removable metal or enamel sheet *h*. Such a sheet is a great advantage, as it aids considerably in keeping the stove clean.

Some gas stoves are provided with a *pilot*, which is a tiny flame of gas that is controlled by a button on the gas pipe to which the stop-cocks are attached. The pilot is kept lighted, and when it is desired to light a burner, pressing the button causes the flame to shoot near enough to each burner to ignite the gas. However, whether the burners are lighted in this way or by applying a lighted match, they should never be lighted until heat is required; likewise, in order to save gas, they should be turned off as soon as the cooking is completed.

To produce the best results, the flame given off by gas should be blue. A flame that is yellow and a burner that makes a noise when lighted, indicate that the gas flame has caught in the pipe, and to remedy this the gas must be turned out and relighted. When the gas flame coming from a new burner is yellow, it may be taken for granted that not enough air is being admitted to make the proper mixture. To permit of the proper mixture, each gas pipe extending from the stop-cock and terminating in the burner is provided with what is called a *mixer*. This device consists of several slots that may be opened or closed by turning part *a*, thus making it a simple matter to admit the right amount of air to produce the desired blue flame. If burners that have been in use for some time give off a yellow flame, it is probable that the trouble is caused by a deposit of soot or burned material. Such burners should be removed, boiled in a solution of washing soda or lye until the holes in the top are thoroughly cleaned, and then replaced and adjusted. As long as the flame remains yellow, the gas is not giving off as much heat as it should produce and is liable to smoke cooking utensils black. Therefore, to get the best results the burners should be thoroughly cleaned every now and then in the manner mentioned. Likewise, the pan beneath the burners, which may be removed, should be cleaned very frequently, and the entire stove should be wiped each time it is used, for the better such a stove is taken care of, the better will it continue to do its work.

## Fireless-cooking Gas Stoves

A style of gas stove that meets with favor in many homes is the so-called fireless-cooking gas stove. Such a stove has the combined advantages of a fireless cooker, which is explained later, and a gas stove, for it permits of quick cooking with direct heat, as well as slow cooking with heat that is retained in an insulated chamber, that is, one that is sufficiently covered to prevent heat from escaping. In construction, this type of stove is similar to any other gas stove, except that its oven is insulated and it is provided with one or more compartments for fireless cooking, as at *a* and *b*. Each of these compartments is so arranged that it may be moved up and down on an upright rod, near the base of which, resting on a solid plate *c*, is a gas burner *d*, over which the insulated hood of the compartment fits.

When it is desired to cook food in one of these compartments, the hood is raised, as at *b*, and the gas burner is lighted. The food in the cooker is allowed to cook over the lighted burner until sufficient heat has been retained or the process has been carried sufficiently far to permit the cooking to continue without fire. Then the insulated hood is lowered until the compartment is in the position of the one shown at *a*. It is not necessary to turn off the gas, as this is done automatically when the hood is lowered.

## KEROSENE STOVES AND THEIR OPERATION

As has been mentioned, kerosene is used considerably as a fuel in localities where gas cannot be obtained. Kerosene stoves are not unlike gas stoves, but, as a rule, instead of having built-in ovens, they are provided with portable ovens, which are heated by placing them on top of the stove, over the burners. Such stoves are of two types, those in which cotton wicks are used, as in oil lamps, and those which are wickless, the former being generally considered more convenient and satisfactory than the latter.

Oil for the burners, or lamps, *a* is stored in the container *b*, which may be of glass or metal, and it is supplied to the reservoir of each burner by the pipe *c*. Each burner is provided with a door *d*, which is opened when it is desired to light the wick. The flame of each burner is controlled by the screw *e*, which serves to raise or lower the wick, and the heat passes up to the opening *f* in the top of the stove through the cylindrical pipe above the burner. The arrangement of a wickless kerosene stove is much the same as the one just described, but it is so constructed that the oil, which is also stored in a tank at the side, flows into what is called a burner bowl and burns from this bowl up through a perforated chimney, the quantity of oil used being regulated by a valve attached to each bowl.

The burners of kerosene stoves are lighted by applying a match, just as the burners of a gas stove are lighted. In some stoves, especially those of the wickless type, the burners are so constructed that the flame can rise to only a certain height. This is a good feature, as it prevents the flame from gradually creeping up and smoking, a common occurrence in an oil stove. The kerosene-stove flame that gives the most heat, consumes the least fuel, and produces the least soot and odor is blue in color. A yellow flame, which is given off in some stoves, produces more or less soot and consequently makes it harder to keep the stove clean. Glass containers are better than metal containers, because the water that is always present in small quantities in kerosene is apt to rust the metal container and cause it to leak. To prevent the accumulation of dirt, as well as the disagreeable odor usually present when an oil stove is used, the burners should be removed frequently and boiled in a solution of washing soda; also, if a wick is used, the charred portion should be rubbed from it, but not cut, as cutting is liable to make it give off an uneven flame.

## ELECTRIC STOVES AND UTENSILS

### Electric Stoves

Electric stoves for cooking have been perfected to such an extent that they are a great convenience, and in places where the cost of electricity does not greatly exceed that of gas they are used considerably. In appearance, electric stoves are very similar to gas stoves which illustrates an electric stove of the usual type. The oven *a* is located at one side and contains a broiler pan *b*. On top of this stove are openings for cooking, into which fit lids *c* that have the appearance of ordinary stove lids, but are in reality electrical heating units, called hotplates. Heat for cooking is supplied by a current of electricity that passes through the hotplates, as well as through similar devices in the oven, the stove being connected to the supply of electricity at the connection-box *d*, which is here shown with the cover removed.

The heat of the different hotplates and the oven is controlled by several switches *e* at the front of the stove. Each of these switches provides three degrees of heat—high, medium, and low—

and just the amount of heat required for cooking can be supplied by turning the switch to the right point. Below the switches are several fuse plugs *f* that contain the fuses, which are devices used in electrical apparatus to avoid injury to it in case the current of electricity becomes too great. It is not absolutely necessary to have flue connections for an electric stove, as such a stove does not require a draft and gives off no products of combustion to be carried away. In fact, one of the favorable points about an electric stove is that it produces no dirt and causes no inconvenience. When the cooking is done, the electricity can be turned off, after which the stove quickly cools. When electricity is used for cooking, cooking utensils, methods, and recipes can be applied in the same ways as when other means of producing heat are employed.

## Small Electric Utensils

In addition to electric stoves, there are a number of smaller electrical cooking utensils that can be attached to an electric-light socket or a wall socket. Among these are percolators, toasters, hotplates, or grills, chafing dishes, egg poachers, and similar devices. An idea of such utensils for cooking, which shows an electric toaster which shows a hotplate, or grill. The toaster is arranged so that bread to be toasted may be placed on each side, as well as on top, of an upright part that gives off heat when the current of electricity is turned on. The grill is so constructed that a pan for cooking may be placed under and on top of the part that gives off heat.

## ESSENTIALS OF COOKERY (PART 1) EXAMINATION QUESTIONS

(1) Give in its full sense the meaning of the term cookery.

(2) How may the housewife control the cost of her foods?

(3) (*a*) Explain the difference between waste and refuse. (*b*) To what is leakage in the household due?

(4) What three important matters enter into the problem of purchasing food?

(5) (*a*) Name the five substances that are found in food, (*b*) Of what value is a knowledge of these food substances?

(6) (*a*) What is the function of protein in the body? (*b*) Mention the principal sources of protein, (*c*) Explain the effect of heat on foods that contain protein.

(7) (*a*) With what do carbohydrates supply the body? (*b*) Mention the two forms of carbohydrates and also some of the foods in which each may be found.

(8) What is a calorie?

(9) Give five reasons for cooking food.

(10) Mention the twelve principal processes employed in the cooking of food.

(11) Describe one method of cooking with: (*a*) dry heat; (*b*) moist heat; (*c*) hot fat.

(12) (*a*) At what temperature does water boil? (*b*) How is hard water affected by boiling? (*c*) Explain the uses of water in cooking.

(13) (*a*) What generally controls the kind of stove to be used for cooking? (*b*) Explain how it is possible to keep down the cost of cooking in using fuel.

(14) Mention the best way in which to become familiar with the operation of a stove.

(15) (*a*) Of what value is gas as a fuel? (*b*) What kind of gas flame is best for cooking?

(16) Suppose that a gas meter registers 72,500 cubic feet on March 1, and that on April 1 the hand of the left dial is between 7 and 8, that of the middle dial is between 5 and 6, and that of the right dial is at 5. At 90 cents a 1,000 cubic feet, what is the cost of the gas consumed?

(17) (*a*) How is heat produced in a stove? (*b*) What is the purpose of the dampers of a stove?

(18) (*a*) How should the dampers of a coal range be adjusted so as to heat the oven for baking? (*b*) How should they be adjusted for cooking on top of the stove?

(19) (*a*) What is the purpose of a mixer on a gas stove? (*b*) How may a gas stove be kept in good condition?

(20) How may the burners of a kerosene stove be kept clean?

## UTENSILS FOR COOKING

### Importance of Utensils

While success in cooking, as has been pointed out, depends to a considerable extent on the selection of materials and the proper cooking methods, as well as on an understanding of the stove and fuel employed, the importance of the utensils that are to be used must not be overlooked. As is well known, each cooking utensil is fitted to its particular use; in fact, the wrong kind of pan, dish, or other utensil will not bring about the same result as the right one. This does not mean, however, that the housewife must possess a large supply of every kind of utensil, for, really, the expert cook is known by the small number of utensils she uses. Of course, the proper handling of utensils, as well as the right selection of them, will come with experience, but before she starts to cook the beginner should endeavor to plan definitely what must be provided. She should likewise remember that the use of an unnecessary number of utensils not only will increase the labor involved in preparing a dish, but will affect considerably the amount of work required to clear them away and wash them after the cooking is done.

The materials of which cooking utensils are made, as well as their shape and size, have also a great bearing on the success with which cooking may be done. As no one material is suitable for all utensils, they are made of various materials, such as wood, tin, glass, enamel, aluminum, sheet iron, and earthenware. In the purchase of a utensil, therefore, it is well to have in mind the use to which the utensil will be put, and then to select one that is made of durable material, that can be easily cleaned, and that will not affect the food that is cooked in it. Likewise, the shape of the utensil should receive consideration, for much depends on it. To be satisfactory, a utensil should be without seams or curved edges, because it is difficult to remove particles of food that collect in such places. A vessel that is hard to wash should be avoided, and one that will tip easily is not desirable, either.

The size of utensils must be determined by the number of persons for whom food is to be cooked, for the amount of food to be prepared indicates whether a large or a small utensil should be selected. On the other hand, the length of time required for foods to cook depends to a large extent on the size and shape of the utensil. When food is to be cooked a long time, a deep vessel with a comparatively small surface exposed for evaporation should be chosen; but for quick cooking, use should be made of a shallow utensil that will allow a great deal of surface to be exposed, as the evaporation will be accomplished more rapidly. In furnishing a kitchen, it is well to begin with a few essential utensils of the best quality that can be obtained, and then, as needed, to add other well-selected utensils to the equipment.

## MATERIALS USED FOR UTENSILS

### Aluminum

Because of the properties of aluminum, this metal is used extensively for cooking utensils. It is more costly than most of the materials employed for this purpose, but while the first cost of aluminum pans and kettles may seem large, the extra expense is justified by the durability of the utensils.

They last much longer than utensils made of many other materials, for when aluminum is hammered and rolled it becomes extremely hard. Some aluminum utensils are very thin, and since they melt and dent very easily they are suitable for only light, careful handling. Although heavier aluminum utensils are more expensive than the lighter ones on account of the metal required and the manufacturing process involved, they are harder and more durable.

Cast aluminum is used for large vessels, such as those required in institutions where large quantities of food are cooked and where pots and kettles are subjected to extremely hard wear, but this is the most expensive kind, for in order to make the aluminum hard enough for casting some harder metal must be mixed with it. One of the disadvantages of aluminum is that it is not always easy to clean, but this is overbalanced by the fact that foods do not burn so readily in aluminum utensils as in other kinds, since the heat is evenly distributed by this metal.

### Enamel

Good enamel cooking utensils are desirable for some purposes and are only moderately expensive. Utensils made of enamel are not so durable as those made of metal, because excessive heat or a sharp blow will cause the enamel to chip. Enamel utensils come in various colors, and all can be kept clean easily, but the gray enamel is considered to be the best for wear.

### Iron and Steel

Utensils made of iron and steel are usually inexpensive, but some, especially those of iron, are heavy. These metals are used principally for such utensils as frying pans, or skillets, griddles, waffle irons, and kettles for deep-fat frying. Sheet iron makes excellent shallow pans for baking cookies and other cakes, very satisfactory bread pans, and the best kind of pans for omelet and other frying.

### Tin

The cheapest metal from which cooking utensils are made is tin, but it is not generally used for utensils in which food is to be cooked, because it melts at too low a temperature. Tin is used, however, for such small articles as measures, cutters, apple corers, sieves, strainers, and other things of this kind, and it is especially desirable for them.

### Copper

Before iron was known copper was the principal material for cooking utensils. The chief point in favor of copper is its durability, but utensils made of it are not practical for use in the ordinary kitchen because they are expensive, heavy, and very difficult to keep clean.

### Glass

Utensils made of heavy glassware are much used for cooking. Glass utensils are especially desirable for custards and other dishes that the cook likes to watch while cooking or that are to be served in the baking dish. Glass cooking utensils possess the advantage of retaining the heat well.

### Wood

Certain utensils made of wood are required in a cooking outfit, a molding board of hardwood and a smaller wooden cutting board being particularly necessary in every kitchen. Bowls in which to chop foods, rolling pins, and mixing spoons are usually made of hardwood, and when such wood is used for them they are entirely satisfactory.

## LABOR-SAVING DEVICES

A LABOR-SAVING DEVICE is any apparatus that will permit a certain piece of work to be accomplished with less exertion than would be necessary to do the same thing without it. A sink and a dustpan are labor-saving devices just as truly as are a bread mixer and a vacuum cleaner, but because a sink and a dustpan are necessities as well, they are not usually thought of as true labor-saving devices. The newer appliances for saving labor are often considered to be quite unnecessary, and indeed some of them are. It is only when such apparatus will, with less labor involved and less time consumed in the process, secure results as good as or better than will another device, and when the cleaning and care of it do not consume so much time and labor as is saved by using it, that it may be considered a true labor-saving device. Each housewife must decide for herself whether the expense of a so-called labor-saving device is greater than the value of the time and strength she would use without such a device.

## Common Labor-saving Devices

Every housewife does not have occasion to use all the devices that have been invented to save labor, but a number of these are in such common use, produce such good results, and save so much time and effort that they should be found in every kitchen. Among them is the *rotary egg beater* (*a*). This is so made that one revolution of the wheel to which the crank is attached does about five times as much work as can be done with a fork or with an *egg whip*, which is shown in (*b*). Another inexpensive device that is a real help is the *potato ricer.*

This device is really a press through which any fruit or vegetable can be put to make a purée. It is used considerably for mashing potatoes, as it makes them perfectly smooth and saves considerable time and labor. Still another useful device is the *meat chopper*, or *grinder*, which is shown in Fig. 3. Such a device clamped to the edge of a table takes the place of a chopping bowl and knife, and in addition to being more sanitary it permits the work to be done in a shorter time and with less effort. Besides the devices mentioned, there are many small labor-saving devices, such as the *apple corer*, the *berry huller*, the *mayonnaise mixer*, etc., the merits of which every busy housewife will do well to consider.

## Bread and Cake Mixers

Where baking is done for only a small number of persons, bread and cake mixers are not indispensable, but they save much labor where baking is done on a large scale. It is comparatively easy, for instance, to knead dough for three or four loaves of bread, but the process becomes rather difficult when enough dough for eight to sixteen loaves must be handled. For large quantities of bread and cake, mixers, when properly used, are labor-saving. In addition, such devices are sanitary, and for this reason they are used in many homes where the bakings are comparatively small.

The type of bread mixer in common use is shown in Fig. 4. It consists of a covered tin pail *a* that may be fastened to the edge of a table by the clamp *b*. Inside of the pail is a kneading prong *c*, in the shape of a gooseneck, that is revolved by turning the handle *d*. The flour and other materials for the dough are put into the pail, and they are mixed and kneaded mechanically by turning the handle.

A cake mixer, the usual type of which is shown in Fig. 5, is similar in construction to a bread mixer. Instead of a pail, however, for the dough ingredients, it has a deep pan *a*, and instead of one kneading prong it has several prongs, which are attached to two arms *b*, as shown. These arms are revolved by gear-wheels *c* that fit in a large gearwheel *d* attached to a shaft *e*, which is turned by means of a handle *f*. The large number of mixing prongs in a cake mixer are necessary, because cake dough must be thoroughly stirred and beaten, whereas in bread making the dough must be made to form a compact mass.

## Dish-washing Machines

Although machines for washing dishes are to be had, they are most helpful where large numbers of people are served and, consequently, where great quantities of dishes are to be washed. Such machines are usually large and therefore take up more space than the ordinary kitchen can afford. Likewise the care and cleaning of them require more labor than the washing of dishes for a small family entails. Large quantities of hot water are needed to operate mechanical dish washers, and even where they are installed, the glassware, silver, and cooking utensils must, as a rule, be washed by hand.

## Fireless Cooker

A device that has proved to be really labor-saving is the fireless cooker. It consists of an insulated box *a* lined with metal and divided into compartments *b*, with pans *c* that fit into them. Hotplates, or stones, as they are sometimes called, are frequently used if the article to be cooked requires them. These stones, which are shown at *d*, are supported in the compartments by metal racks *e*, and they are lifted in and out by means of wire handles *f*.

To use a fireless cooker properly, the food must be cooked for a short time on the stove; then it must be tightly covered and placed in one of the insulated compartments. If hotplates are to be used they must be heated in the same manner. The food loses its heat so gradually in the fireless cooker that the cooking proceeds slowly but effectually. When the previous heating has been sufficient, the food will be cooked and still warm when the cooker is opened hours later. Some articles of food occasionally need reheating during the process. By this method of cooking there is no loss of flavor or food value, and the food usually requires no further attention after being placed in the cooker. It also permits of economy in both fuel and time.

## UTENSILS FOR FURNISHING A KITCHEN

As a guide in purchasing equipment for a kitchen, a list of utensils is here presented. This list is divided into utensils that are necessary and those that are convenient and only at times necessary. In any case, however, the number of utensils and the size must be determined by the quantity of food that is to be prepared.

### Necessary Equipment

- Baking dish with cover Bread box Bread knife
- Bread pans
- Can opener
- Cake knife
- Chopping bowl and knife or food chopper
- Coffee mill
- Coffee pot
- Colander
- Cookie cutter
- Corer, Apple
- Cutting board
- Dishpan
- Double boiler
- Egg beater
- Flour sifter
- Forks

- Frying pan, large
- Frying pan, small
- Garbage can
- Grater
- Kettle covers
- Kettles, two or more
- Knife sharpener
- Knives
- Lemon squeezer
- Long-handled fork
- Measuring cup
- Meat board
- Meat knife
- Mixing bowls
- Mixing spoons
- Molding board
- Muffin pan
- Paring knife
- Pepper shaker
- Pie pans
- Potato masher
- Rinsing, or draining, pan
- Roasting pan
- Rolling pin
- Salt box
- Saucepans
- Spatula
- Tablespoons
- Teakettle
- Teapot
- Teaspoons
- Toaster
- Wire strainer
- Wooden spoon

## UTENSILS FOR FURNISHING A KITCHEN

### Convenient Equipment

- Bread mixer
- Cake coolers
- Cake mixer
- Cake turner
- Casseroles
- Clock
- Coffee percolator
- Containers for spices and dry groceries
- Cookie sheets
- Cream whip Egg whip
- Fireless cooker
- Frying kettle and basket
- Funnel Glass jars for canning
- Griddle
- Ice-cream freezer
- Ice pick
- Jelly molds
- Nest of bowls
- Pan for baking fish
- Potato knife
- Potato ricer
- Ramekins
- Quart measure
- Scales
- Scissors
- Set of skewers
- Steamer
- Waffle iron
- Wheel cart

# 3

# Getting Foods Ready for Cooking

## PRELIMINARY PREPARATION

Before foods that require cooking are cooked or before foods that are to be eaten raw are served, they must be properly prepared, for their palatability and their value as food depend considerably on the way in which they are made ready for cooking or for eating. Of course, the way in which food should be prepared will depend on how it is to be served, but in any event all foods, for the sake of cleanliness, must first be washed with water or wiped with a clean, damp cloth.

The ways in which vegetables and fruits are made ready for cooking vary. Sometimes such foods are cooked with the skins on, and sometimes certain vegetables, such as new potatoes, young carrots and parsnips, vegetable oysters, etc., are made ready in an economical way by scraping off their skins with a knife. Vegetables are also peeled, and when this is done a very sharp knife with a thin blade should be used and as little of the food removed as possible. Still another way of removing the skins of such foods as tomatoes, nuts, and some fruits is by *blanching*. In this process, the skins are loosened so that they may be removed easily, either by immersing the foods in boiling water or by pouring boiling water over them and allowing them to stand in the water for a few minutes, but not long enough to soften them. Blanching used in this sense should not be confused with the same word when it means "to take color out" and has reference to a process of bleaching. Only when the word means "to remove the covering of" can it be applied to the peeling of tomatoes, fruits, and nuts. Vegetables and fruits may be cooked whole or they may be cut into chunks, or pieces, or into slices.

In order to get meats ready for cooking, it is necessary to wipe them clean and usually to trim off all unnecessary bone, fat, and skin. Meats may be cooked in large pieces or small pieces or they may be ground, depending on the cooking process to be used. Before cooking poultry and fish, they should be thoroughly cleaned and then trimmed and cut to suit the cooking process chosen. If desired, the bones may be removed from poultry or fish before cooking, and sometimes it is advantageous to do so. Cream and raw eggs may be whipped or beaten light before they are served or cooked, and after such foods as fruits, vegetables, meats, and fish have been cooked, they may be sliced, chopped, ground, mashed, or cut into dice, or small pieces.

## MIXING OF FOOD INGREDIENTS

### Processes Involved in Mixing

In cookery, the mixing of ingredients is done for several purposes—to produce a certain texture, to give a smoothness or creaminess to a mixture, or to impart lightness. Various processes are involved in the mixing of ingredients, and the results that are accomplished depend entirely on the method that is selected. The most important of these processes with brief explanations of what they mean follow.

BEATING is a rapid motion that picks up material from the bottom and mixes it with that nearer the surface. It is done with a spoon, a fork, an egg whip, or, if the mixture is thin, with a rotary egg beater. Sometimes beating is done for the purpose of incorporating air and thus making the mixture light.

STIRRING is usually done with a spoon, and is accomplished by moving the spoon in circles, around and around, through ingredients contained in a pan or a bowl. This is the method that is generally applied to the simple mixing of ingredients.

FOLDING is a careful process whereby beaten egg or whipped cream is added to a mixture without destroying its lightness. It is accomplished by placing the egg or cream on top of a mixture in a bowl or a pan, and then passing a spoon down through both and bringing up a spoonful of the mixture and placing it on top. This motion is repeated until the two are well blended, but this result should be accomplished with as few strokes as possible.

RUBBING is done by pressing materials against the side of

a bowl with the back of a spoon. This is the process that is applied when butter and other fats are to be mixed with such dry ingredients as sugar and flour.

CREAMING consists in continuing the rubbing process until the texture becomes soft and smooth and is of a creamy consistency.

CUTTING-IN is a method used to combine butter with flour when it is desired to have the butter remain hard or in small pieces. It is done by chopping the butter into the flour with a knife.

SIFTING is shaking or stirring material through a sifter having a fine wire mesh. It is done to remove foreign or coarse material, to impart lightness, or to mix dry ingredients together.

RICING is a process whereby certain cooked foods, such as fruits, vegetables, meats, and fish, may be reduced to the form of a purée. This result is accomplished by forcing the cooked material through a ricer.

## Application of Mixing Processes

In applying the various mixing processes, it is well to bear in mind that good results depend considerably on the order of mixing, as well as on the deftness and thoroughness with which each process is performed. This fact is clearly demonstrated in a cake in which the butter and sugar have not been actually creamed, for such a cake will not have the same texture as one in which the creaming has been done properly. It is also shown in angel food or sunshine cake, for the success of such a cake depends largely on the skill employed in folding in the whites of eggs or in beating the yolks. On the other hand, the lightness of pastry and the tenderness of cookies depend on how each is rolled out, and the kneading of bread is a process that demonstrates that many things can be learned by actually doing them.

As progress is made with these cookery lessons, therefore, the application of the mixing processes should not be overlooked. Beginners in cookery, owing possibly to the fact that at first they cannot handle soft material skilfully, are liable to make the mistake of getting the ingredients too stiff. Yet no beginner need feel the least bit discouraged, for ability in this direction comes with experience; indeed, just as skill in sewing, embroidering, and other processes comes about by practice and persistent effort, so will come skill in cooking.

## MEASURING

Uniform results in cookery depend on accurate measurement. Of course, there are some cooks—and good ones, too—who claim that they do not measure, but as a matter of fact they have, through long experience, developed a judgment, or "sense," of measurement, which amounts to the same thing as if they actually did measure. Still, even these cooks cannot be absolutely sure of securing as satisfactory results time after time as are likely to follow the employment of a more accurate method. Therefore, to secure the best results, every kitchen should be supplied with the proper measuring utensils, which are scales, a measuring cup, and a set of measuring spoons, or a standard tablespoon and a standard teaspoon.

### Scales

The material to be weighed is placed on the platform at the top, and the weight of it is indicated on the dial by a pointer, or hand. Sometimes these scales are provided with a scoop in which loose materials may be placed in weighing. Such scales furnish a correct means not only of measuring materials, but of verifying the weights of foods from the market, the butcher shop, or the grocery. To use them properly, the housewife should learn to balance them exactly, and when she is weighing articles she should always allow for the weight of the container or receptacle, even if it is only the paper that holds the food.

### Measuring Cups

Weighing the articles called for in a recipe is often a less convenient method than measuring; therefore, in the preparation of foods, measuring is more often resorted to than weighing. As accuracy in measurement is productive of the best results, it is necessary that all measures be as accurate and definite as possible. For measuring the ingredients called for in recipes, use is generally made of a measuring cup like that shown in Fig. 8. Such a cup is designed to hold 2 gills, or 1/2 pint, and it is marked to indicate thirds and quarters, so that it may be used for recipes of all kinds. If a liquid is to be measured with such a cup, it should be filled to the brim, but if dry material is to be measured with it, the material should be heaped up in the cup with a spoon and then scraped level with a knife, in the manner shown in Fig. 9. In case fractions or parts of a cup are to be measured, the cup should be placed level and stationary and then filled evenly to the mark indicated on the cup itself.

Many times it will be found more convenient to measure dry materials with a spoon. This can be done with accuracy if it is remembered that 16 tablespoonfuls make 1 cup, or 1/2 pint; 12 tablespoonfuls, 3/4 cup; 8 tablespoonfuls, 1/2 cup; and 4 tablespoonfuls, 1/4 cup. If no measuring cup like the one just described is at hand, one that will hold 16 level tablespoonfuls of dry material may be selected from the kitchen supply of dishes. Such a cup, however, cannot be used successfully in measuring a half, thirds, or fourths; for such measurements it will be better to use a spoon.

As a rule, it will be found very convenient to have two measuring cups of standard size, one for measuring dry ingredients and the other for measuring moist or wet ones. If it is impossible to have more than one, the dry materials should be measured first in working out a recipe, and the fats and liquids afterwards. Whatever plan of measuring is followed, however, it should always be remembered that recipes are written for the definite quantities indicated and mean *standard*, not approximate, cupfuls, tablespoonfuls, and teaspoonfuls.

## Measuring Spoons

In addition to a measuring cup or two, a set of measuring spoons will be found extremely convenient in a kitchen. However, if it is impossible to obtain such a set, a teaspoon and a tablespoon of standard size will answer for measuring purposes. Three level teaspoonfuls are equal to 1 tablespoonful. When a spoon is used, it is heaped with the dry material and then leveled with a knife.

Precautions to Observe in Measuring.—In measuring some of the materials used in the preparation of foods, certain points concerning them should receive attention. For instance, all powdered materials, such as flour, must first be sifted, as the amount increases upon sifting, it being definitely known that a cupful of unsifted flour will measure about 1-1/4 cupfuls after it is sifted. Lumps, such as those which form in salt and sugar, should be thoroughly crushed before measuring; if this is not done, accurate measurements cannot be secured, because lumps of such ingredients are more compact than the loose material.

Butter and other fats should be tightly packed into the measure, and if the fat is to be melted in order to carry out a recipe, it should be melted before it is measured. Anything measured in a cup should be poured into the cup; that is, the cup should not be filled by dipping it into the material nor by drawing it through the material.

## Tables of Weights and Measures

As foods are sold by weight and by measure, and as recipes always call for certain weights and measures, it is absolutely necessary that every person engaged in the purchase and preparation of foods should be familiar with the tables of weights and measures in common use for such purposes in the United States and practically all other English-speaking countries. In addition, it will be well to have a knowledge of relative weights and measures, so as to be in a position to use these tables to the best advantage.

The table used ordinarily for weighing foods is the table of AVOIRDUPOIS WEIGHT. Another table of weights, called the table of *Troy weight*, is used by goldsmiths and jewelers for weighing precious metals. It should not be confused with avoirdupois weight, however, because its pound contains only 12 ounces, whereas the avoirdupois pound contains 16 ounces. The table of avoirdupois weight, together with the abbreviations of the terms used in it, is as follows:

| AVOIRDUPOIS WEIGHT | | |
|---|---|---|
| 437-1/2 grains (gr.) | = 1 ounce | oz. |
| 16 ounces | = 1 pound | lb. |
| 100 pounds | = 1 hundredweight | cwt |
| 20 hundredweight | = 1 ton | T. |

### *2,000 pounds*

Although 2,000 pounds make 1 ton, it is well to note that 2,240 pounds make 1 *long ton* (L.T.). The long ton is used by coal dealers in some localities, but the ton, sometimes called the *short ton,* is in more general use and is the one meant unless long ton is specified.

The table of LIQUID MEASURE is used for measuring all liquids, and is extremely useful to the housewife. This table, together with the abbreviations of its terms, is as follows:

| LIQUID MEASURE | | |
|---|---|---|
| 4 gills (gi.) | = 1 pint | pt. |
| 2 pints | = 1 quart | qt. |
| 4 quarts | = 1 gallon | gal |
| 31-1/2 gallons | = 1 barrel | . |
| 2 barrels | = 1 hogshead | bbl |
| 63 gallons | | . |

The table of DRY MEASURE is used for measuring dry foods, such as potatoes, dried peas and beans, etc. The table of dry measure, with its abbreviations, follows:

| DRY MEASURE | | |
|---|---|---|
| 2 pints (pt.) | = 1 quart | qt |
| 8 quarts | = 1 peck | . |
| 4 pecks | = 1 bushel | p |

Tables of RELATIVE WEIGHTS AND MEASURES are of value to the housewife in that they will assist her greatly in coming to an understanding of the relation that some of the different weights and measures bear to one another. For example, as dry foods are sold by the pound in some localities, it will be well for her to know the approximate equivalent in pounds of a definite quantity of another measure, say a quart or a bushel of a certain food.

Likewise, she ought to know that when a recipe calls for a cupful it means 1/2 pint, as has been explained. Every one is familiar with the old saying, "A pint's a pound the world around," which, like many old sayings, is not strictly true, for while 1 pint is equal to 1 pound of some things, it is not of others. The following tables give approximately the relative weights and measures of most of the common foods:

| APPROXIMATE MEASURE OF 1 POUND OF FOOD | |
|---|---|
| Beans, dried | 2 CUPFULS |
| Butter | 2 CUPFULS |
| Coffee, whole | 4 CUPFULS |
| Corn meal | 3 CUPFULS |
| Flour | 4 CUPFULS |
| Milk | 2 CUPFULS |
| Molasses | 1-1/2 CUPFULS |
| Meat, chopped, finely packed | 2 CUPFULS |
| Nuts, shelled | 3 CUPFULS |
| Oats, rolled | 4 CUPFULS |
| Olive oil | 2-1/2 CUPFULS |
| Peas, split | 2 CUPFULS |
| Raisins | 3 CUPFULS |
| Rice | 2 CUPFULS |
| Sugar, brown | 2-2/3 CUPFULS |
| Sugar, granulated | 2 CUPFULS |
| Sugar, powdered | 2-3/4 CUPFULS |

*Approximateweight of 1tablespoonfulof Food*

| | |
|---|---|
| Butter | 1/2 OUNCE |
| Corn starch | 3/8 OUNCE |
| Flour | 1/4 OUNCE |
| Milk | 1/2 OUNCE |
| Sugar | 1/2 OUNCE |

*Approximateweight of 1 Cupfulof Food*

| | |
|---|---|
| Butter | 8 OUNCES |
| Corn meal | 5 OUNCES |
| Corn starch | 6 OUNCES |
| Flour | 4 OUNCES |

| | |
|---|---|
| Milk | 8 OUNCES |
| Molasses | 10 OUNCES |
| Nuts, shelled | 4 OUNCES |
| Raisins | 5 OUNCES |
| Sugar | 8 OUNCES |

In measuring, you will find the following relative proportions of great assistance:

3 tsp. = 1 Tb.

16 Tb. = 1 c.

ABBREVIATIONS OF MEASURES.—In order to simplify directions and recipes in books relating to cookery, it is customary to use the abbreviations of some weights and measures. Those which occur most frequently in cook books are the following:

tsp. for teaspoonful

pt. for pint

Tb. for tablespoonful

qt. for quart

c. for cupful

oz. for ounce

lb. for pound

## ORDER OF WORK

For successful results in cookery, the work to be done should be planned beforehand and then carried on with systematic care. By following such a plan, a waste of time and material will be prevented and good results will be secured, for there will be little chance for mistakes to occur. The order of work here outlined will serve to make clear the way in which cooking processes can be carried out satisfactorily.

First, read the quantity and kind of ingredients listed in the recipe, and study carefully the method by which they are to be prepared and combined. In so doing, determine whether the dish is too expensive and whether the amounts called for will make a dish sufficient in size for the number of persons to be served. If they are too large, carefully divide them to make the right quantity; if they are too small, multiply them to make them enough.

The heat itself, which plays such an important part in cooking, should receive attention at the proper time. If the fuel to be used is coal or wood and baking is to be done, build the fire long enough before it is needed, so that it will be burning evenly and steadily.

Then, while the recipe is being prepared, provided it is to be baked, regulate the heat of the oven. If gas or kerosene is to be used, light it after the recipe is read, and regulate it during the measuring and mixing of the ingredients.

Before proceeding to prepare a dish, clear enough working space for the utensils that are to be used, as well as for carrying on the various operations without feeling crowded. Then, on the cleared space, place the necessary measuring utensils, such as a measuring cup, a knife, a teaspoon, and a tablespoon. Select a bowl or a pan for mixing, a spoon for stirring, and, when needed, an egg whip or beater for eggs and separate bowls in which to beat them. Choose the utensil in which the mixture is to be cooked, and, if necessary, grease it. During the process of preparing the dish, measure accurately all the ingredients to be used, and check them up with the recipe, so as to be sure that none are missing and that each one is in its proper amount.

If all these steps are accurately taken, the mixing, which is the next step, can be accomplished quickly and without error. With all the ingredients properly combined, the mixture is ready for the last step, the cooking or the baking. This must be done with the utmost care, or an otherwise properly prepared dish may be spoiled.

## TABLE FOR COOKING FOODS

So that the beginner in cookery may form a definite idea of the length of time required to cook certain foods, there is presented here what is commonly known as a *cookery time table.* It should be remembered that the time required to cook food is influenced by many factors. For instance, the age of vegetables and fruits very largely determines how long they should be cooked; tough meats and fowl require longer cooking than tender ones; and the heat of the oven has much to do with the length of time required for cooking, especially the process of baking or roasting Therefore, while this time table will prove of great help to beginners, it can serve only as a guide. To determine whether or not foods have been cooked long enough, it is advisable to apply the proper tests, which are given later in discussing the various foods rather than to depend solely on the time table. In this table, the length of time for cooking is given in minutes (abbreviated min.) and hours (abbreviated hr.)

## COOKERY TIME TABLE

### *Meats and Fish*

#### Broiled

| | |
|---|---|
| Bacon | 3 to 5 min. |
| Chicken | 20 to 25 min. |
| Fish | 15 to 20 min. |
| Fish, slices | 10 to 15 min. |
| Fish, very small | 5 to 10 min. |
| Lamb chops | 6 to 8 min. |
| Quail or squabs | 8 to 10 min. |
| Steak, thick | 10 to 15 min. |
| Steak, thin | 5 to 7 min. |
| Veal chops | 6 to 10 min. |

#### Boiled

| | |
|---|---|
| Beef, corned | 3 to 4 hr. |
| Chicken, 3 lb | 1 to 1-1/4 hr. |
| Fish, bluefish, cod, or bass, 4 tc 5 lb | 20 to 30 min. |
| Fish, slices, 2 to 3 lb | 20 to 25 min. |
| Fish, small | 10 to 15 min. |
| Fowl, 4 to 5 lb | 2 to 3 hr. |
| Ham, 12 to 14 lb | 4 to 5 hr. |
| Mutton, leg of | 2 to 3 hr. |
| Tongue | 3 to 4 hr. |

#### *Roasted*

| | |
|---|---|
| Beef, rib or loin, 5 lb., rare | 1 hr. 5 min. |
| Beef, rib or loin, 5 lb., well done | 1 hr. 20 min. |
| Beef, rib or loin, 10 lb., rare | 1 hr. 30 min. |

| | |
|---|---|
| Beef, rib or loin, 10 lb., well done | 2 hr. |
| Beef, rump, 10 lb., rare | 1 hr. 30 min. |
| Beef, rump, 10 lb., well done | 2 hr. |
| Chicken, 4 or 5 lb | 1-1/2 to 2 hr. |
| Duck, 5 to 6 lb | 1-1/4 to 1-1/2 hr. |
| Fish, 3 to 5 lb | 45 to 60 min. |
| Fish, small | 20 to 30 min. |
| Goose, 10 lb | 2 to 2-1/2 hr. |
| Lamb, leg of | 1-1/4 to 1-3/4 hr. |
| Mutton, saddle | 1-1/4 to 1-1/2 hr. |
| Pork, rib, 5 lb | 2 to 2-1/2 hr. |
| Turkey, 10 lb | 2-1/2 to 3 hr. |

*Vegetables*

*Boiled*

| | |
|---|---|
| Asparagus | 20 to 30 min. |
| Beans, lima or shell | 40 to 60 min. |
| Beans, string | 30 to 45 min. |
| Beets, old | 4 to 6 hr. |
| Beets, young | 45 to 60 min. |
| Brussels sprouts | 15 to 25 min. |
| Cabbage | 35 to 60 min. |
| Carrots | 3/4 to 2 hr. |
| Cauliflower | 20 to 30 min. |
| Green corn | 8 to 12 min. |
| Macaroni | 30 to 40 min. |
| Onions | 45 to 60 min. |
| Peas | 25 to 60 min. |
| Potatoes | 30 to 45 min. |
| Rice | 20 to 30 min. |
| Spinach | 20 to 30 min. |
| Turnips | 1/2 to 1-1/2 hr. |
| Vegetable oysters | 3/4 to 1-1/2 hr. |

### *Baked Foods*

| | |
|---|---|
| Beans | 6 to 8 hr. |
| Biscuits, baking powder | 15 to 25 min. |
| Biscuits, yeast | 10 to 25 min. |
| Bread, ginger | 20 to 30 min. |
| Bread, loaf | 40 to 60 min. |
| Cake, corn | 20 to 30 min. |
| Cake, fruit | 1-1/4 to 2 hr. |
| Cake, layer | 15 to 20 min. |
| Cake, loaf | 40 to 60 min. |
| Cake, pound | 1-1/4 to 1-1/2 hr. |
| Cake, sponge | 45 to 60 min. |
| Cookies | 6 to 10 min. |
| Custard | 20 to 45 min. |
| Muffins, baking powder | 15 to 25 min. |
| Pastry | 30 to 45 min. |
| Potatoes | 45 to 60 min. |
| Pudding, Indian | 2 to 3 hr. |
| Pudding, rice (poor man's). | 2 to 3 hr. |

## CARE OF FOOD

### Reasons for Care

Although, as has been explained, the selection and preparation of foods require much consideration from the housewife who desires to get good results in cookery, there is still one thing to which she must give attention if she would keep down the cost of living, and that is the care of food. Unless food is properly taken care of before it is cooked, as well as after it is cooked—that is, the left-overs—considerable loss is liable to result through its spoiling or decaying. Both uncooked and cooked food may be kept wholesome in several ways, but before these are discussed it may be well to look into the causes of spoiling. With these causes understood, the methods of caring for foods will be better appreciated, and the results in buying, storing, and handling foods will be more satisfactory.

To come to a knowledge of why foods spoil, it will be well to note that nature abounds in *micro-organisms*, or living things so minute as to be invisible to the naked eye. These micro-organisms are known to science as *microbes* and *germs*, and they are comprised of *bacteria*, *yeasts*, and *molds*, a knowledge of which is of the utmost importance to the physician and the farmer, as well as the housewife. Just in what ways these are beneficial to the farmer and the physician is beyond the scope of the subject of cookery, but in the household their influence is felt in three ways: They are the cause of the decay and spoiling of foods; they are of value in the preparation of certain foods; and they are the cause of contagious diseases. It will thus be seen that while some microbes are undesirable, others exert a beneficial action.

It is only within comparatively recent years that the action of micro-organisms has been understood. It is now definitely known that these minute living things seize every possible chance to attack articles of food and produce the changes known as fermentation, putrefaction, souring, and decay. Micro-organisms that cause fermentation are necessary in bread making and vinegar making, but they are destructive to other foods, as, for example, those which are canned or preserved. Organisms that cause putrefaction are needed in the making of sauer kraut, salt rising bread, and cheese. Molds also help to make cheese, but neither these nor putrefactive organisms are desirable for foods other than those mentioned. It should be remembered, however, that even those foods which require micro-organisms in their making are constantly in danger of the attacks of these small living things, for unless something is done to retard their growth they will cause food to sour or decay and thus become unfit for consumption.

Some foods, of course, withstand the attacks of micro-organisms for longer periods of time than others. For example, most fruits that are protected by an unbroken skin will, under the right conditions, keep for long periods of time, but berries, on account of having less protective covering, spoil much more quickly. Likewise, vegetables without skins decay faster than those with skins, because they have no protective covering and contain more water, in which, as is definitely known, most micro-organisms thrive.

If food is to be kept from decaying, the housewife must endeavor to prevent the growth of micro-organisms, and she can best accomplish this if she is familiar with the ways in which they work. It is for this reason that, whether she possesses a scientific knowledge of bacteria or not, an understanding of some practical facts concerning why food spoils and how to keep it from decaying is imperative. In this part of cookery, as in every other phase, it is the reason why things should be done that makes all that relates to the cooking of food so interesting. In all parts of the work there are scientific facts underlying the processes, and the more the housewife learns about these, the more she can exercise the art of cookery, which, like all other arts, depends on scientific principles.

## METHODS OF CARE CLASSIFICATION

As has been pointed out, it is not the mere presence of micro-organisms that causes the spoiling of food, but their constant growth. Therefore, to keep milk from souring, meat from spoiling, bread from molding, canned fruit from fermenting, and so on, it is necessary to know what will prevent the growth of these minute organisms. Different foods require different treatment. Some foods must be kept very cold, some must be heated or cooked, others must be dried, and to others must be added preservatives. An unwarrantable prejudice has been raised in the minds of many persons against the use of preservatives, but this is due to the fact that the term is not properly understood. In this use, it means anything that helps to preserve or keep safe the food to which it is added. Sugar, salt, spices, and vinegar are all preservatives, and are added to food as much for the purpose of preserving it as for seasoning it.

## CANNING AND DRYING OF FOODS

Among the common methods of caring for foods that are to be used at a future time are canning and drying. CANNING, which is discussed fully in another Section, consists in preserving sterile foods in sealed cans or jars. The aim in canning is to prevent the growth of micro-organisms, and to do this the process known as *sterilizing*—that is, the destroying of bacteria and other micro-organisms by means of heat—is resorted to. Canning theories are different now from what they were in former times. For example, housewives formerly made heavy, rich preserves of available fruits because it was thought that sugar must be used in large quantities in order to keep or prevent them from spoiling. While it is true that the sugar assisted, science has since proved that sterilizing is what must be done, so that now only the sugar desired for sweetening need be used.

The other method of keeping food, namely, DRYING, depends for its success on the fact that such micro-organisms as bacteria cannot grow unless they have a considerable quantity of moisture or water. Molds grow on cheese, bread, damp cloth or paper, or articles that contain only a small amount of moisture, but bacteria need from 20 to 30 per cent. of water in food in order to grow and multiply. This explains why in high altitudes and dry climates foods keep for a long time without artificial means of preservation. It also explains why the old-fashioned housekeeper dried fruits and why the preservation of certain meats is accomplished by the combined methods of smoking and drying, the creosote of the smoke given off from the wood used in this process acting as a preservative. All the grains, which are very dry, keep for long periods of time, even centuries, if they are protected from the moisture of the air. Peas, beans, and lentils, as well as dried biscuits and crackers, are all examples of how well food will keep when little or no moisture is present.

KEEPING FOODS WITH ICE

Although, as has just been pointed out, moisture is required for the growth of some micro-organisms, both moisture and warmth are necessary for the growth of most of the organisms that cause molding, putrefaction, and fermentation. It is definitely known, also, that in winter or in cold climates food can be kept for long periods of time without any apparent change; in fact, the lower the temperature the less likely are foods to spoil, although freezing renders many of them unfit for use. These facts are what led up to the scientific truth that keeping foods dry and at a low temperature is an effective and convenient method of preventing them from spoiling and to the invention of the refrigerator and other devices and methods for the cold storage of foods.

## The Refrigerator

For home use, the refrigerator offers the most convenient means of keeping foods in good condition. As is well known, it is a device that, by means of air cooled by the melting of ice or in some other manner, keeps food at a temperature near the freezing point. All refrigerators are constructed in a similar manner, having two or more layers of wood between which is placed an insulating material, such as cork, asbestos, or mineral wool. The food compartments are lined with tile, zinc, or other rust-proof material, and the ice compartment is usually lined with rust-proof metal, so as to be water-tight and unbreakable. Any refrigerator may be made to serve the purpose of preserving food effectively if it is well constructed, the ice chamber kept as full of ice as possible, and the housewife knows how to arrange the foods in the food chambers to the best advantage.

The construction and use of refrigerators are based on the well-known scientific fact that air expands and rises when it becomes warm. This can be proved by testing the air near the ceiling of a room, for no matter how warm it is near the floor it will always be warmer above. The same thing occurs in a refrigerator. As air comes in contact with the ice, it is cooled and falls, and the warm air is forced up. Thus the air is kept in constant motion, or circulation.

In such refrigerators, there is usually a small food compartment directly under the ice chamber, and this is the coldest place in the refrigerator. Here should be stored the foods that need special care or that absorb odors and flavors readily, such as milk, butter, cream, meat, etc., because at this place the air, which circulates in the manner indicated by the arrow, is the purest. The foods that give off odors strong enough to taint others should be kept on the upper shelves of the refrigerator, through which the current of air passes last before being freed from odors by passing over the ice.

This type is so built as to produce on each side a current of air that passes down from the ice at the center and back up to the ice near the outside walls, as shown by the arrows. A different arrangement is required for the food in this kind of refrigerator, those which give off odors and flavors being placed in the bottom compartment, or farthest from the ice, and those which take up odors and flavors, on the top shelf, or nearest the ice.

## Care of Food in Refrigerator

The proper placing of foods in a refrigerator is extremely important, but certain precautions should be taken with regard to the food itself. Cooked foods should never be placed in the refrigerator without first allowing them to cool, for the steam given off when a dish of hot food comes in contact with the cold air makes the refrigerator damp and causes an undue waste of ice by warming the air. All dishes containing food should be wiped dry and carefully covered before they are placed in the refrigerator, so as to keep unnecessary moisture out of it. As butter and milk are likely to become contaminated with odors given off by other foods, they should be properly protected if there is not a separate compartment in which to keep them. The milk bottles should always be closed and the butter carefully wrapped or put in a covered receptacle. Onions, cabbage, and other foods with strong odors, when placed in the refrigerator, should be kept in tightly closed jars or dishes, so that the odors will not escape. Before fresh fruits and perishable vegetables—that is, vegetables that decay easily—are put into the refrigerator, they should be carefully looked over and all decayed portions removed from them. No food should be placed in the ice chamber, because this will cause the ice to melt unnecessarily.

### Care of the Refrigerator

It is essential that all parts of the refrigerator be kept scrupulously clean and as dry as possible. To accomplish this, nothing should be allowed to spoil in it, and anything spilled in the refrigerator should be cleaned out immediately. The foods that are left over should be carefully inspected every day, and anything not likely to be used within a day or so should be disposed of. At least once a week the food should be removed from all compartments, the racks taken out, the drain pipe disconnected, and each part thoroughly washed, rinsed with boiling water, and dried. The inside of the refrigerator should likewise be washed, rinsed, and wiped dry, after which the drain pipe should be connected, the shelves put back in place, and the food replaced.

The ice chamber of the refrigerator should also be cleaned frequently, the best time to do this being when the ice has melted enough to be lifted out conveniently. To prevent the ice from melting rapidly when it is out of the refrigerator, it may be wrapped in paper or a piece of old blanket, but this covering must be removed when the ice is replaced in the chamber, in order to allow the ice to melt in the refrigerator. Otherwise, it would be impossible to chill the refrigerator properly, the temperature remaining the same as that outside, for it is as the ice gradually melts that the air in the refrigerator becomes cool. Of course, every effort should be made to keep the ice from wasting. Therefore, while the refrigerator should be kept in a convenient place, it should not be exposed to too great heat; also, the doors should be kept tightly closed, and, as has already been explained, hot foods should not be put in until they are sufficiently cooled. Attention must be given to the care of the refrigerator, for only when it is clean and dry can the growth of bacteria that attack foods be prevented.

## KEEPING FOODS WITHOUT ICE

While a refrigerator simplifies the preserving of cooked foods and those subject to quick decay, there are many communities in which it is not possible to procure ice conveniently, thus making it necessary to adopt some other means of keeping food. Then, too, there are generally quantities of foods, such as winter vegetables, apples, etc., that cannot be stored in a refrigerator, but must be taken care of properly. In such cases, the method of storing depends to a certain extent on conditions. On many farms there are spring houses in which foods may be stored in order to keep them cool during very warm weather; but in the majority of homes, the cellar, on account of its being cool, is utilized for the storage of large quantities of food and even for keeping the more perishable foods when ice cannot be obtained.

### Storing Foods in Cellars

In order that a cellar may furnish a safe place for keeping food, it must be well built and properly cared for. If it is dug in wet ground and is not well drained, it will become musty and damp, and fruits and vegetables stored in it will be attacked by mold. A small part of the cellar should be without a floor, as many winter vegetables seem to keep better when placed on dry ground, but the remainder should have a flooring of either well-matched boards or cement that can be kept clean and dry. Ventilation must also be supplied; otherwise, odors will be retained that will taint the food kept in the cellar. To allow the passage of air and light from the outside and thus secure proper ventilation, the cellar should be provided with windows. These will also assist very much in the cleaning and airing of the cellar, processes that should never be overlooked if good results are desired. In addition to the cleaning of the cellar, constant attention should be given to the foods kept there. Foods that have spoiled or are beginning to spoil should be disposed of quickly, for decayed food that is not removed from the cellar will affect the conditions for keeping other foods and may be injurious to the health of the family.

All foods likely to be contaminated by dust and flies in the cellar must be carefully covered. A screened frame fastened to the wall with brackets is excellent for this purpose, because it prevents the attack of vermin and permits of ventilation. If canned goods are to be stored, a cellar cupboard is a very good place in which to keep them. Separate bins should, if possible, be provided for fruits, potatoes, and other winter vegetables, such bins should be so built as to allow air to pass through them.

### Window Boxes

The woman who lives in an apartment where there is no cellar and who does not wish to keep ice in the refrigerator through the winter will find a window box a very good device in which to keep food. Such a box is also a convenience for the woman who has a cellar, but wishes to save steps. A box of this kind is built to fit a kitchen or a pantry window, and is placed outside of the window, so that the opening comes toward the room. Such an arrangement will make the contents of the box easily accessible when the window is raised. A box for this purpose may be made of wood or galvanized iron, and it is usually supported by suitable brackets. Its capacity may be increased by building a shelf in it half way to the top, and provided it is made of wood, it can be more easily cleaned if it is lined with table oilcloth.

## STORING OF NON-PERISHABLE FOODS

It may seem unnecessary to give much attention to the storing of foods that do not spoil easily, but there are good reasons why such foods require careful storage. They should be properly cared for to prevent the loss of flavor by exposure to the air, to prevent the absorption of moisture, which produces a favorable opportunity for the growth of molds, and to prevent the attacks of insects and vermin. The best way in which to care for such foods is to store them in tightly closed vessels. Earthenware and glass jars, lard pails, coffee and cocoa cans, all carefully cleaned and having lids to fit, prove to be very satisfactory receptacles for such purposes.

Unless coffee, tea, cocoa, spices, and prepared cereals are bought in cans or moisture-proof containers, they should be emptied from the original packages and placed in jars that can be tightly closed, so that they will not deteriorate by being exposed to the air or moisture. For convenience and economy, these jars or cans should be labeled. Sugar and salt absorb moisture and form lumps when exposed to the air, and they, too, should be properly kept. A tin receptacle is the best kind for sugar, but for salt an earthenware or glass vessel should be used. It is not advisable to put these foods or any others into cupboards in paper bags, because foods kept in this way make disorderly looking shelves and are easily accessible to vermin, which are always attracted to food whenever it is not well protected.

Canned goods bought in tin cans do not need very careful storage. It is sufficient to keep them in a place dry enough to prevent the cans from rusting. Foods canned in glass, however, should be kept where they are not exposed to the light, as they will become more or less discolored unless they are stored in dark places.

Flour, meals, and cereals stored in quantities develop mold unless they are kept very dry. For the storing of these foods, therefore, wooden bins or metal-lined boxes kept in a dry place are the most satisfactory.

## STORING OF SEMIPERISHABLE FOODS

Practically all vegetables and fruits with skins may be regarded as semiperishable foods, and while they do not spoil so easily as some foods, they require a certain amount of care. Potatoes are easily kept from spoiling if they are placed in a cool, dry, dark place, such as a cellar, a bin furnishing a very good means for such storage. It is, of course, economical to buy potatoes in large quantities, but if they must be kept under conditions that will permit them to sprout, shrivel, rot, or freeze, it is better to buy only a small quantity at a time. Sweet potatoes may be bought in considerable quantity and kept for some time if they are wrapped separately in pieces of paper and packed so that they do not touch one another.

Carrots, turnips, beets, and parsnips can be kept through the winter in very much the same manner as potatoes. They deteriorate less, however, if they are covered with earth or sand. Sometimes, especially in country districts, such winter vegetables are buried in the ground out of doors, being placed at a depth that renders them safe from the attacks of frost. Cabbage will keep very well if placed in barrels or boxes, but for long keeping, the roots should not be removed. Pumpkin and squash thoroughly matured do not spoil readily if they are stored in a dry place.

Apples and pears may be stored in boxes or barrels, but very fine varieties of these fruits should be wrapped separately in paper. All fruit should be looked over occasionally, and those which show signs of spoiling should be removed.

## MENUS AND RECIPES

As practically every woman knows, a MENU, or *bill of fare,* consists of a certain number of dishes given in the order in which they are to be served; likewise, she knows that the dishes called for in a menu must be prepared according to a RECIPE, or *receipt,* which is the list of ingredients of a mixture giving the exact proportions to be used, together with proper directions for compounding.

In all good recipes the items are tabulated in the order in which they are needed, so as to save time and produce good results. Items tabulated in this manner also serve to minimize the danger of omitting some of the ingredients of a recipe, for they can be easily checked up when they are given in the proper order.

In preparing recipes, the beginner in cookery usually has difficulty in judging the size of a recipe. The experienced housewife will not follow a recipe exactly when she thinks it will produce more food than she needs to meet the requirements of her family; instead, she will reduce the quantities to suit her wants. Likewise, if a recipe will not provide enough, she will increase the quantities accordingly.

Just how to judge whether or not a recipe will make what is wanted comes only with experience, but the beginner may be guided by the fact that it is never wise to prepare more than enough of one kind of dish, unless, of course, it can be used to good advantage as a left-over. On the other hand, if a recipe is for food that can be kept and used for another meal later, it often pays to make up more, so as to save time, fuel, and labor. In any event, it is always advisable to follow explicitly the directions that are given, for if the recipe is of the right kind they will be given so that success will result from carrying them out in detail.

In order that the beginner in cookery may form a definite idea of the manner in which the dishes of a menu, or bill of fare, may be prepared so that they will be ready to serve in their proper order at meal time, there is here given a simple dinner menu, together with the recipes for preparing the dishes called for and the order in which they should be prepared. While these recipes are not intended to teach methods of cookery, which are taken up later, the student is advised to prepare the menu for her own satisfaction and so that she will be able to report on the success she has had with each dish.

## MENU

Pan-Broiled Chops Mashed Potatoes Creamed Peas Cabbage Salad Orange Fluff with Sauce

## RECIPES

### Pan-broiled Chops

Buy the necessary number of pork, veal, or lamb chops, and proceed to cook them according to the directions previously given for pan broiling. Season with salt and pepper just before removing the chops from the pan.

### Mashed Potatoes

Peel the desired number of potatoes, put to cook in a sufficient amount of boiling salted water to cover well, and cook until the potatoes are tender enough to be easily pierced with a fork. Remove from the fire and drain off the water. Mash the potatoes with a wooden or a wire potato masher, being careful to reduce all the particles to a pulpy mass in order to prevent lumps, or put them through a ricer. When sufficiently mashed, season with additional salt, a dash of pepper, and a small piece of butter, and add hot milk until they are thinned to a mushy consistency, but not too soft to stand up well when dropped from a spoon. Then beat the potatoes vigorously with a large spoon until they are light and fluffy.

### Creamed Peas

Boil until they are soft, two cupfuls of fresh peas in 1 quart of water to which have been added 1 tablespoonful of salt and 2 of sugar, and then drain; or, use 1 can of peas, heat them to the boiling point in their liquid, and then drain. A part of the water in which the fresh peas were cooked or the liquid on the canned peas may be used with an equal amount of milk to make a sauce for the peas, or all milk may be used.

### Sauce For Peas

1 c. of milk, or 1/2 c. liquid from peas and 1/2 c. milk 1 Tb. butter 1/2 tsp. salt 1 Tb. flour

Melt the butter in a saucepan or a double boiler, work in the flour and salt until a smooth paste is formed, and add the liquid that has been heated. Stir until thick and smooth. Add to the peas, reheat, and serve.

### Cabbage Salad

1/2 medium-sized head of cabbage

1/2 tsp. salt

1 small red or green sweet pepper

Dash of pepper

1 small onion

Salad dressing

Shred the cabbage finely by cutting across the leaves with a sharp knife or a cabbage shredder. Chop the pepper and onion into very small pieces and add to the cabbage. Mix well and add the salt and pepper.

### Cabbage-salad Dressing

3/4 c. vinegar

1/2 tsp. mustard, if desired

1/4 c. water

1/2 tsp. salt

2 Tb. butter

3 Tb. sugar 1 Tb. flour

Heat the water and the vinegar; melt the butter in a saucepan, add to it the flour, mustard, salt, and sugar, stir until well blended, and then pour in the hot liquid. Cook for a few minutes, stirring constantly to prevent the formation of lumps. Pour over the cabbage while hot; allow it to cool and then serve on plates garnished with lettuce.

### Orange Fluff

1/2 c. sugar

1/4 c. orange juice

5 Tb. corn starch

1 Tb. lemon juice Pinch of salt

2 egg whites 1 pt. boiling water

Mix the corn starch and sugar and salt, stir into the boiling water, and cook directly over the fire until the mixture thickens. Continue to cook, stirring constantly for 10 minutes, or place in a double boiler and cook 1/2 hour. Beat the egg whites until they are stiff.

When the corn starch is cooked, remove from the fire and mix thoroughly with the fruit juices. Pour over the beaten egg whites and stir slightly until the eggs and corn starch are mixed. Pour into sherbet glasses or molds wet with cold water and set aside until ready to serve.

### Sauce for Orange Fluff

1 Tb. corn starch 3/4 c. boiling water

2 Tb. butter 3/4 c. sugar 2 egg yolks 1/4 c. orange juice 1 Tb. lemon juice

Moisten the corn starch with a little cold water and stir in 1/2 cupful of the boiling water. Cook for 10 or 15 minutes. Cream the butter, add the sugar and egg yolks, beat the mixture with a fork, and add the remaining 1/4 cupful of boiling water. Stir this into the corn starch and cook until the eggs thicken slightly. Remove from the fire and add the orange and lemon juices. Serve cold over the orange fluff.

In the preparation of a meal, it is impossible to follow the order of service given in a menu, because of the different lengths of time required to prepare the different dishes. The order in which the menu here given should be prepared will therefore serve to show the way in which other meals may be planned or other menus carried out. Each recipe for this menu is planned to serve six persons, but it can be easily changed in case a different number are to be served. For instance, if there are only four in the family, two-thirds of each ingredient should be used; and if only three, just one-half of each. If eight are to be served, one-third will have to be added to each of the amounts. As has been pointed out, just a little thought will show how other numbers may be provided for.

In preparing the foods called for in this menu, the dessert, which is the last thing given, should be prepared first, because time must be allowed for it to cool before serving. In fact, it may be prepared a half day before it is to be served. So as to allow sufficient time to mash the potatoes after they have boiled, they should be made ready to put on the stove about 3/4 hour before the meal is to be served. After the potatoes have been put on to boil, the peas, provided fresh ones are to be used, should be put on to cook, and then the sauce for them should be made. If canned peas are to be used, the sauce should be made after the potatoes have been put on the stove and the peas should be heated and combined with the sauce just before broiling the chops. The cabbage salad may then be prepared, and put in a cool place until it is to be served. The chops should be broiled last, because it is necessary that they be served immediately upon being taken from the fire.

# 4

# Cereals

## PRODUCTION, COMPOSITION, AND SELECTION

### Production of Cereals

*Cereals,* which is the term applied to the edible seeds of certain grains, originated with the civilization of man. When man lived in a savage state, he wandered about from place to place and depended for his food on hunting and fishing; but as he ceased his roaming and began to settle in regions that he found attractive, it was not long before he became aware of the possibilities of the ground about him and realized the advantage of tilling the soil as a means of procuring food. Indeed, the cultivation of the soil for the production of food may be considered as one of the first steps in his civilization. Among the foods he cultivated were grains, and from the earliest times to the present day they have been the main crop and have formed the chief food of people wherever it is possible to produce them.

The grains belong to the family of grasses, and through cultivation their seeds, which store the nourishment for the growth of new plants, have been made to store a sufficient amount of nourishment to permit man to collect and use it as food. The name cereals was derived from the goddess Ceres, whom the Romans believed to be the protector of their crops and harvests. Numerous grains are produced, but only eight of these cereals are used extensively as food, namely, wheat, corn, oats, rice, barley, rye, buckwheat, and millet.

### Abundance of Production

With the exception of the desert lands and the Arctic regions, cereals of some kind are grown over the entire world. Some varieties thrive in the hot countries, others flourish in the temperate regions, and still others mature and ripen in the short warm season of the colder northern climates. In fact, there is practically no kind of soil that will not produce a crop of some variety of grain. Since grains are so easily grown and are so plentiful, cereals and foods made from them furnish a large part of the world's food supply. Indeed, about one-fourth of all the food eaten by the inhabitants of the world, when it is considered as a whole, is made up of cereals.

## Economic Value of Cereals

The abundance of the world's grain supply makes the cost so moderate that many of the poorer classes of people in various countries, especially those in the Far East, live almost entirely on cereals. Still there is another factor that controls the low cost of cereals and grains and keeps them within the means of all classes of people, and that is their excellent keeping quality. They require very little care and will keep for an indefinite period of time. Because of their unperishable nature, they may be stored in large quantities and distributed to consumers as they are needed and at a price that is fairly uniform.

Since the cost of cereals is moderate, they should form a large proportion of the diet of the entire family, especially if the family's income will allow only a limited sum to be spent for food. Some cereals, of course, are much cheaper than others, and in purchasing this kind of food the housewife should be governed accordingly. Those which require an elaborate manufacturing process in their preparation for the market are the most expensive, but they have an advantage in that they require practically no preparation before serving. For the varieties that must be cooked, the cost of preparing the dish, especially if the price of fuel is high, must be taken into consideration, for unless some thought is given to the economical use of the fuel, as well as to the method of cooking employed, the cost of the prepared dish may be greatly increased. However, in the preparation of cereals, very little skill or energy is required and a general knowledge of the best methods for one of them can, as a rule, be applied to all.

## Cereal Products

Besides the cereals already mentioned, a number of products of cereals are extensively used in cookery, chief among them being flour, corn starch, and other starches. Although every housewife should possess knowledge of the uses of each of these, instruction in them is not given until later.

This Section includes particularly the study of grains—whole, cracked, flaked, and those made into grits or meal—and the use and the serving of them, as well as ready-to-eat cereals, which are commonly referred to as *breakfast foods*. The only additional foods to which attention is given at this time are macaroni, spaghetti, and foods of a similar nature, for as these are made from wheat they are truly cereal products. In their preparation for the table, the rules that govern the other cereal foods apply also in a large measure to them.

## COMPOSITION OF CEREALS

The composition of all cereals is similar, yet each one has its distinguishing feature. While all the five food substances—water, mineral matter, protein, fat, and carbohydrate—are to be found in cereals, they occur in different quantities in the various kinds. Some contain large quantities of protein and others practically none, and while certain ones have considerable fat others possess comparatively small quantities. A characteristic of all cereals, however, is that they contain a large amount of carbohydrate and a small amount of water. It is well to remember, though, that while the food substances of cereals are found in sufficient quantities to sustain life, they will not permit a person to live for long periods of time exclusively on this form of food.

Likewise, it will be well to observe that the foods made from a certain grain will be quite similar in composition to the grain itself; that is, any change in the composition of the foods must be brought about by the addition of other substances.

All grains are similar in general structure, too. The largest proportion of carbohydrate lies in the center, this substance growing less toward the outside of the grain. The protein lies near the outside, and grows less toward the center. Fat is found in small amounts scattered through the entire grain, but most of it is found in the *germ*, which is a tiny portion of the grain from which the new plant sprouts. The mineral matter of cereals is found chiefly just inside the bran, or outer covering, so that when this covering is removed, as in the process of preparation for food, a certain amount of mineral matter is generally lost.

## Protein in Cereals

The cereals are essentially a carbohydrate food, but some also yield a large proportion of protein. In this respect they differ from the animal foods that produce the principal supply of protein for the diet, for these, with the exception of milk, do not yield carbohydrates. The grain that contains the most protein is wheat, and in the form in which protein occurs in this cereal it is called *gluten,* a substance that is responsible for the hardness of wheat. The gluten, when the wheat is mixed with water or some other liquid, becomes gummy and elastic, a fact that accounts for the rubbery consistency of bread dough. Cereals that contain no gluten do not make bread successfully. Next to wheat, rye contains protein in the greatest amount, and rice contains the least. Although protein is the most expensive of the food substances, the kind of protein found in cereals is one of the cheaper varieties.

## Fat in Cereals

The fat of cereals helps to contribute to their heat-and energy-producing qualities, and, besides, it is one of the cheaper sources of this food substance. Of the eight grains, or cereals, used as food, oats and corn contain the most fat, or heat-producing material. The oil of corn, because of its lack of flavor, is frequently used in the manufacture of salad oil, cooking oil, and pastry fat. The fat that occurs in cereals becomes rancid if they are not carefully stored. In the making of white flour, the germ of the wheat is removed, and since most of the fat is taken out with the germ, white flour keeps much better than graham flour, from which the germ is not abstracted in the milling process.

## Carbohydrate in Cereals

The food substance found in the greatest proportion in cereals is carbohydrate in the form of starch. Cereals contain many times more starch than any of the other food substances, rice, which is fully three-fourths starch, containing the most, and oats, which are less than one-half starch, the least. Starch is distributed throughout the grain in tiny granules visible only under the microscope, each being surrounded by a covering of material that is almost indigestible. In the various grains, these tiny granules differ from one another in appearance, but not to any great extent in general structure, nutritive value, or digestibility, provided they are cooked thoroughly. The large amount of carbohydrate, or starch, in cereals explains why they are not hard to digest, for, as is well known, starch is more easily digested than either protein or fat. This and the fact that some grains contain also a large amount of fat account for the high energy-producing quality of cereals. While it is safe to say that cereals are chiefly valuable for their starch, the tissue-building material in some grains, although in small proportion, is in sufficient quantity to place them with the protein foods.

## Mineral Matter in Cereals

Cereals contain seven or eight of the minerals required in the diet. Such a variety of minerals is sure to be valuable to the human body, as it is about one-half of the whole number required by the body for its maintenance. Since, as has already been explained, much of the mineral matter lies directly under the coarse outside covering, some of it is lost when this covering is removed. For this reason, the grains that remain whole and the cereal products that contain the entire grain are much more valuable from the standpoint of minerals than those in which the bran covering is not retained. If a sufficient percentage of minerals is secured in the diet from vegetables, fruits, and milk, it is perhaps unnecessary to include whole cereals; but if the diet is at all limited, it is advisable to select those cereals which retain the original composition of the grain.

## Water in Cereals

Cereals contain very little water in their composition. This absence of water is a distinct advantage, for it makes their nutritive value proportionately high and improves their keeping quality. Just as the strength of a beverage is lowered by the addition of water, so the nutritive value of foods decreases when they contain a large amount of water. On the other hand, the keeping quality of cereals could scarcely be improved, since the germs that cause foods to spoil grow only in the presence of water. This low proportion of water also permits them to be stored compactly, whereas if water occurred in large amounts it would add materially to their bulk.

## Cellulose in Cereals

In addition to the five food substances that are found in all cereals, there is always present another material known as cellulose, which, as is pointed out elsewhere, is an indigestible material that occurs on the outside of all grains, as the bran covering, and covers the starch granules throughout the inside of the grain. In fact, it forms a sort of skeleton upon which the grains are built. As long as the cellulose remains unbroken, it prevents the grain from being digested to any extent.

However, it forms a valuable protective covering for the grain and it has a certain value, as bulk, in the diet, a fact that is ignored by some persons and overrated by others. It is well to include at least some cellulose in cereal foods when they are taken in the diet, because its presence tends to make food less concentrated.

## Table Showing Composition of Cereals

Not all grains, or cereals, contain the same amount of food substances and cellulose; that is, while one may be high in protein it may be lacking in some other food substance. The relation that the various grains bear to one another with regard to the food substances and cellulose is clearly set forth in Table I. In this table, under the various food substances and cellulose, the grains, with the exception of millet, are mentioned in the order of their value, ranging from the highest down to the lowest in each of the food substances and cellulose.

Thus, as will be seen, wheat is highest in protein and rice is lowest, oats are highest in fat and rye is lowest, and so on. Also, as will be observed, while wheat is highest in protein, it is, as compared with the other cereals, sixth in fat, fourth in carbohydrate, fourth in cellulose, and fifth in mineral matter. In this way may be compared all the other cereals to see in just what way they are of value as a food.

*Table I: Composition Of Cereals*

| Protein | Fat | Carbohydrate | Cellulose | Mineral Matter or Ash |
|---|---|---|---|---|
| Wheat | Oats | Rice | Oats | Oats |
| Rye | Corn | Rye | Buckwheat | Barley |
| Oats | Barley | Corn | Barley | Buckwheat |
| Barley | Buckwheat | Wheat | Wheat | Rye |
| Corn | Rice | Barley | Rye | Wheat |
| Buckwheat | Wheat | Buckwheat | Corn | Corn |
| Rice | Rye | Oats | Rice | Rice |

## CEREALS AS A FOOD

### Uses Of Cereals

Cereals and cereal products play a very important part in the food problem, for the prosperity of a country depends on its grain crops and the people of all classes are dependent on them for food. This is evident when it is known that they form a greater proportion of the food consumed than any other single food material. In their widespread consumption, they have many and varied uses. In truth, a meal is seldom served without some cereal food, for if no other is used, bread of some description is almost always included. Besides bread, a cooked or a dry cereal is usually served for breakfast, and for some persons this constitutes the main breakfast dish, providing a nourishing and easily digested food when served with milk or cream. This food is especially desirable for children, and for this reason is always among the first solid foods fed to them.

While to most persons the word cereal suggests the idea of a breakfast food, because cereals are used most often for that purpose, they find their place in other meals than breakfast. Although they are used less often on the dinner table than elsewhere, they frequently have an important place there, for a number of them are commonly used as dinner dishes and others might be used more frequently, and to advantage, too. In this connection, they are used in soups, and in certain forms, usually the whole or slightly crushed grain, they take the place of a vegetable. Some of them, particularly rice, are often used with meat or cheese in making an entree or in combination with eggs, milk, fruit, or various flavorings as a dessert to be served with a heavy or a light meal. Cold cooked cereal is often sliced and sautéd and then served with meat or some other heavy protein dish. Cereals are also used for lunch or supper, perhaps more often than for dinner, and because of their easy digestion they are to be recommended for the evening meal for all members of the family, but especially for children. When used in this way, they may be served with cream, as for breakfast, or prepared in any other suitable way. Whenever cereals are served, whether alone or in combination with other foods, the result is an economical dish and usually an easily digested one, unless, of course, the food with which they are combined is expensive or indigestible. But, to whatever use cereals are put, unless they are thoroughly cooked they are not easily digested and they lose much of their value. In fact, the ready-to-eat cereals, which have been thoroughly cooked, are preferable to those which are poorly cooked in the home.

## SELECTION AND CARE OF CEREALS

So that the housewife may go about the selection of cereals in an intelligent manner, it may be well for her to know how they are prepared for market. After the grains are harvested, the first step in their preparation consists in thrashing, which removes the husks from the outside.

In some countries, thrashing is done entirely by hand, but usually it is accomplished by machinery of a simple or a more elaborate kind. Occasionally no further treatment is applied, the whole grains being used as food, but generally they receive further preparation. Sometimes they are crushed coarsely with or without the bran covering, and in this form they are known as *grits.* At other times they are ground finer and called *meal,* and still finer and called *flour,* being used mostly in these two forms for the making of various kinds of breads.

Then, again, grains are rolled and crushed, as, for example, *cracked wheat* and *rolled oats.*

Various elaborate means have been devised by which cereals are prepared in unusual ways for the purpose of varying the diet. Sometimes they are used alone, but often certain other materials are used in their preparation for the market. For example, the popular flake cereals, such as corn flakes, are cooked with salt and sometimes with sugar and then rolled thin. Some of the cereals are thoroughly cooked, while others are malted and toasted, but the treatment to which they are subjected is generally given to them to improve their flavor and to aid in the work of digestion.

### Factors That Govern Cereal Selection

Besides knowing about the ways in which cereals are prepared for market, the housewife should be familiar with the factors that govern their selection for use as food. In the first place, cereals should be chosen to suit the needs and tastes of the members of the family, and then attention should be given to the forms in which they can be purchased. Some cereals are sold in sealed packages, while others can be bought in bulk. Each, however, has its advantages. Those sold loose are often lower in price than those sold in package form, but there is a question as to whether, with the chances for incorrect weight, the bulk foods are really much cheaper. Cleanliness is, of course, of greater importance with cereals that do not require cooking than with those which are subjected to high temperatures in order to prepare them for the table. Therefore, from the standpoint of cleanliness, there is no advantage in purchasing rice and similar raw cereals in packages.

The next thing to consider in the purchase of cereals is their cost. They vary considerably in price, but it has been determined that in food value there is little difference, pound for pound, between the cheap and the expensive cereals, the variation in price being due to their abundance or scarcity and the method used in preparing them for market. The entirely uncooked ones are the cheapest, the partly cooked ones are medium in price, and the thoroughly cooked ones are the most expensive. This difference, however, is practically made up by the expense of the fuel required to prepare them for the table, the cheapest cereal requiring the most fuel and the most expensive, the least.

Besides varying in price, the different kinds of cereals offer the housewife an opportunity to select the one that is most convenient for her. Those which are ready to serve are the best for the meal to which the least possible amount of time can be given for preparation. The other kinds require cooking, of course, but this need not be a hindrance, for they can be prepared on one day and reheated for breakfast the following day, or they can be cooked overnight by the fireless-cooker method. In the case of such cereals, long cooking is usually necessary for good flavor and easy digestion; consequently, the cooking method that will accomplish the desired result with the least expenditure of fuel is the most economical one and the one to select.

## Table of Grain Products

As a further aid in coming to an understanding of cereals, or grains, and their value, there are given in Table II the various uses to which grains are put and the forms in which they occur as food. In this table, as will be observed, the form of the grain product is mentioned first and then the grain from which it is made. A careful study of this table will be profitable to the housewife.

*Table II: Grain Products*

| | | |
|---|---|---|
| Cereals | Whole Grains | Pearl barley |
| | | Hulled wheat |
| | | Hominy: Corn |
| | | Corn |
| | | Rice |
| | Crushed Grains | Farina: Wheat or corn |
| | | Cream of Wheat: Wheat |
| | | Cracked Wheat: Wheat |
| | | Hominy Grits: Corn |
| | | Wheat Grits: Wheat |
| | | Samp: Corn |
| | Meal | Corn |
| | | Barley |
| | | Rice |
| | | Oats |
| | Prepared Cereals | Flaked: Rye, wheat, rice, corn |
| | | Shredded Grain: Wheat |
| | | Malted Grain: Rye, barley, wheat, and corn |
| | | Puffed Grain: Corn, rice, wheat |
| Starch | Corn | |
| | Rice | |
| | Wheat | |
| Wheat | Macaroni | |
| | Vermicelli | |
| | Spaghetti | |
| Glucose | | |
| Usually corn | | |
| Sirup | | |
| Cereal Coffee | | Wheat |
| | Rye | |
| | Barley | |
| Flour | | |
| Wheat | | |
| Rye | | |
| Corn Buckwheat | | |
| Rice | | |
| Liquors | All grains | |
| Malted Drinks | | |
| Beer | | |
| Whisky | | |
| Alcohol: All grains | | |
| Feed for animals: | | All grains |

## Care of Cereals

As carriers of disease, cereals are a less dangerous food than any other. This characteristic of cereals is due to the fact that the cooking all of them require in some part of their preparation destroys any disease germs that might be present. They are not likely to be adulterated with harmful material, either; and, in addition, the sealed packages in which many of the cereals are put up keep them clean and free from contamination. However, care must be given to both the uncooked and the factory-prepared varieties of this food. The packages containing ready-to-eat cereals should not be allowed to remain open for any length of time if it is desired to keep them fresh and crisp, for they absorb moisture from the air very quickly. If they do become moist, however, drying in the oven will in most cases restore their freshness. If it is necessary to open a single package of prepared cereal and all of the contents cannot be utilized at once, as, for instance, when only one or two persons are to be served with that particular cereal, the best plan is to empty the remainder into cans or jars that are provided with covers. Uncooked cereals, which are used less quickly than the prepared kinds, are often attacked by mice and other vermin, but such an occurrence can be prevented if the cereal is poured into jars or cans that can be kept tightly closed. Considerable care must be given to flour and cereal products purchased in large quantities, for if they are allowed to collect enough moisture, they will become moldy and lose their flavor, and thus be unfit for use. To preserve them well, they should be kept in metal-lined bins or in bins made of carefully matched boards and in a cool, but not damp, place.

# PREPARATION OF CEREALS FOR THE TABLE

## Purpose of Cooking

As the so-called ready-to-eat cereals require practically no further preparation, attention is here given to only those cereals which need additional treatment to prepare them properly for the table. Raw grains cannot be taken into the body, for they are neither appetizing nor digestible. The treatment to which they must be subjected is cooking, for the structure of grains is such that cooking is the only means by which the coverings of the starch granules can be softened and broken to make them digestible. But this is not the only effect produced by cooking; besides making raw cereals digestible, cooking renders them palatable, destroys any bacteria or parasites that might be present, and, by means of its various methods, provides a variety of dishes that would otherwise be very much limited.

## Changes That Cereals Undergo in Cooking

In the process of cooking, cereals undergo a marked change, which can readily be determined by performing a simple experiment. Place an equal amount of flour or corn starch—both cereal products—in two different glasses; mix that in one glass with cold water and that in the other with boiling water. The mixture in which cold water is used will settle in a short time, but if the substance that goes to the bottom is collected and dried it will be found to be exactly the same as it was originally. The mixture in which boiling water is used, however, will not only become a sticky mass, but will remain such; that is, it will never again resume its original form. This experiment proves, then, that grains that come in contact with water at a high temperature, as in cooking, absorb the water and burst their cellulose covering. This bursting frees the granulose, or the contents of the tiny granules, which are deposited in a network of cellulose, and as soon as this occurs it mixes with water and forms what is called soluble starch. Starch in this state is ready for digestion, but in the original, uncooked state only a very small part of it, if any, is digestible.

## Preparation for Cooking Cereals

Before the cooking of cereals is attempted, it is advisable for the sake of convenience to get out all utensils as well as all ingredients that are to be used and arrange them so that they will be within easy reach.

The utensils and ingredients shown, which are suitable for most methods of cooking cereals and particularly for cooking them by the steaming process, consist of a double boiler *a*; a measuring cup *b*, a knife *c*, and spoons *d* and *e*, for measuring; a large spoon *f*, for stirring; a salt container; and a package of cereal. The housewife will be able to tell quickly from a recipe just what ingredients and utensils she will need, and by following the plan here suggested and illustrated she will find that her work can be done systematically and with the least expenditure of time.

## First Steps in the Process of Cooking

While cereals may be cooked in a variety of ways, the first steps in all the processes are practically the same. In the first place, the required amount of water should be brought to the boiling point, for if the water is boiling the cereal will thicken more rapidly and there will be less danger of lumps forming. Then salt should be added to the water in the proportion of 1 teaspoonful to each cupful of cereal. Next, the cereal should be stirred into the boiling salted water slowly enough to prevent it from forming lumps, and then, being constantly stirred, it should be allowed to cook until it thickens. The process up to this point is called *setting* a cereal, or grain. After the cereal is *set*, it may be boiled, steamed, or cooked in the fireless cooker, but the method of cookery selected should be chosen with a view to economy, convenience, and thoroughness. The terms *setting* and *set* should be thoroughly fixed in the mind, so that directions and recipes in which they are used will be readily understood.

## Cooking Cereals by Boiling

Very often the cereal, after it is set, is allowed to cook slowly until it is ready to serve; that is, the method of *boiling* is practiced. This method, however, is not to be recommended, because it is not economical. Cereals cooked in this way require constant watching and stirring, and even then it is difficult to keep them from sticking to the cooking utensil and scorching or becoming pasty on account of the constant motion. Sometimes, to overcome this condition, a large quantity of water is added, as in the boiling of rice; still, as some of this water must be poured off after the cooking is completed, a certain amount of starch and soluble material is lost.

## Cooking Cereals in the Double Boiler

Probably the most satisfactory way in which to cook cereals, so far as thoroughness is concerned, is in a double boiler, one style of which is shown at *a*. This method of cookery is known as *steaming*, or *dry steaming*, and by it the food itself, after it is set, never comes within 6 or 8 degrees of the boiling point. In this method, the cereal is first set in the small, or upper, pan of the double boiler. This pan, which is covered, is placed into the large, or lower, pan, which should contain boiling water, and the cereal is allowed to cook until it is ready to serve. The water in the large pan should be replenished from time to time, for if it is completely evaporated by boiling, the pan will be spoiled and the cereal in the upper pan will burn.

This method of cooking has several advantages that should not be disregarded. Cereals to which it is applied may be partly cooked on one day and the cooking completed the next morning before breakfast, or they may be completely cooked on one day and merely heated before they are served. Then, when cooked at a temperature slightly below the boiling point, the grains remain whole, but become thoroughly softened, because they gradually absorb the water that surrounds them. In addition, the long cooking that is necessary to prepare them at a low temperature develops a delicious flavor, which cannot be obtained by rapid cooking at the boiling point.

## Cooking Cereals in the Fireless Cooker

In a kitchen that is equipped with a fireless cooker, it is advisable to use this utensil for cereals, for cooking them by this method secures the greatest economy of fuel and effort. As in the preceding methods, the cereal is first set in the pan that fits into the cooker compartment. While the cereal is at the boiling point, this pan is covered tightly and placed in the fireless cooker, where it is allowed to remain until the cereal is ready to be served. The heat that the cereal holds when it is placed in the cooker is retained, and this is what cooks it. Therefore, while this method of cooking requires considerable time, it needs neither additional heat nor labor after the cereal is placed in the cooker. In reality, it is an advantageous way in which to cook cereals, since, if they can be set and placed in the cooker in the evening, they will be ready to serve at breakfast time on the following day.

## Cooking Cereals by Dry Heat

An old method of cooking cereals or starchy foods is called *browning,* or *toasting,* and it involves cooking them by dry heat. A thin layer of grain is spread in a shallow pan and this is placed in a slow oven. After the grains have browned slightly, they are stirred, and then they are permitted to brown until an even color is obtained. By this method the flavor of the cereals is developed and their digestibility increased. Since grains keep much better after they have been subjected to the process of toasting, this means is used extensively for preserving grains and cereal foods.

## Points to Observe in Cooking Cereals

In cooking cereals by any method, except browning, or toasting, it is always necessary to use liquid of some kind. The quantity to use, however, varies with the kind of cereal that is to be cooked, whole cereals and those coarsely ground requiring more liquid than those which are crushed or finely ground. If the liquid is to be absorbed completely when the grain is cooked, it should be in the correct proportion to the grain. To be right, cooked cereals should be of the consistency of mush, but not thin enough to pour. Much attention should be given to this matter, for mistakes are difficult to remedy. Cereals that are too thick after they are cooked cannot be readily thinned without becoming lumpy, and those which are too thin cannot be brought to the proper consistency unless the excess of liquid is evaporated by boiling.

*Gruels* are, of course, much thinner than the usual form of cereal. They are made by cooking cereals rapidly in a large quantity of water, and this causes the starch grains to disintegrate, or break into pieces, and mix with the water. The whole mixture is then poured through a sieve, which removes the coarse particles and produces a smooth mass that is thin enough to pour.

The length of time to cook cereals also varies with their kind and form, the coarse ones requiring more time than the fine ones. Because of this fact, it is difficult to say just how much time is required to cook the numerous varieties thoroughly. However, little difficulty will be experienced if it is remembered that cereals should always be allowed to cook until they can be readily crushed between the fingers, but not until they are mushy in consistency.

# INDIAN CORN, OR MAIZE

## Origin, Classification, and Use

The word *corn* has been applied to various grains and is now used in a variety of ways in different countries. In ancient times, barley was called corn, and at the present time, in some countries, the entire year's food crop is referred to by this name. The English apply the name corn to wheat, and the Scotch, to oats. In the

United States, corn is the name applied to the seed of the maize plant, which is a highly developed grass plant that forms the largest single crop of the country. The seeds of this plant grow on a woody cob, and are eaten as a vegetable when they are soft and milky, but as a grain, or cereal, when they are mature. Corn is native to America and was not known in Europe until Columbus took it back with him. However, it did not meet with much favor there, for it was not grown to any great extent until within the last 50 years. Those who took it to Europe gave it the name *Indian corn*, because they had found the Indians of America raising it.

Of the corn grown in the United States, there are three general kinds: field corn, sweet corn, and pop corn. *Field corn*, as a rule, is grown in large quantities and allowed to mature; then it is fed to animals or ground and cooked for the use of man. This corn consists of three varieties, which are distinguished by the color of the grain, one being white, one yellow, and one red. All of them are made into a variety of preparations, but the white and the yellow are used as food for both man and animals, whereas red field corn is used exclusively for animal food. White corn has a mild flavor, but yellow corn is sometimes preferred to it, because foods made from the yellow variety have a more decided flavor. The two principal varieties of field corn, when prepared as cereal food for man, are *hominy* and *corn meal*. *Sweet corn* is not grown in such large quantities as field corn. It is generally used for food before it is mature and is considered as a vegetable. *Pop corn*, when sufficiently dry, swells and bursts upon being heated. It is used more as a confection than as a staple article of food. Therefore, at this time, consideration need be given to only the principal varieties of field-corn products, which, as has just been stated, are hominy and corn meal.

## RECIPES FOR HOMINY AND CORN MEAL

HOMINY is whole corn from which the outside covering has been removed, and for this reason it is high in food value. Corn in this form may be procured as a commercial product, but it may be prepared in the home at less expense. As a commercial product, it is sold dry by the pound or cooked as a canned food. Dry hominy requires long cooking to make it palatable, and this, of course, increases its cost; but even with this additional cost it is cheaper than canned hominy.

Sometimes corn from which the covering has been removed is ground or crushed to form what is called *samp,* or *grits,* and when it is ground still more finely CORN MEAL is produced. Corn meal is made from both white and yellow corn, and is ground more finely in some localities than in others.

It is sold loose by the pound, but it can also be bought in bags or packages of various sizes from 1 pound up. Corn meal should be included in the diet of every economical family, for it yields a large quantity of food at a moderately low cost. If it is prepared well, it is very palatable, and when eaten with milk or cream it is a food that is particularly desirable for children, especially for the evening meal, because of its food value and the fact that it is easily digested.

So that the importance of these corn products may be understood and the products then used to the best advantage in the diet, recipes are here given for preparing hominy in the home, for dishes in which hominy forms the principal part, and for dishes in which corn meal is used. To get the best results from these recipes and thereby become thoroughly familiar with the cooking processes involved, it is recommended that each one be worked out in detail.

This thought applies as well to all recipes given throughout the various Sections. Of course, to prepare each recipe is not compulsory; nevertheless, to learn to cook right means actually to do the work called for by the recipes, not merely once, but from time to time as the food can be utilized to give variety to the daily menus in the home.

## Hominy

Although, as has been mentioned, prepared hominy may be purchased, some housewives prefer to prepare it themselves. Hominy serves as a foundation from which many satisfactory dishes can be made, as it is high in food value and reasonable in cost. This cereal can be used in so many ways that it is advisable to prepare enough at one time to meet the demands of several meals. The following recipe for making hominy should provide 3 quarts of this cereal; however, as is true of other recipes—a point that should be remembered throughout the various lessons—the quantities given may be increased or decreased to meet with the requirements of the household.

(*Sufficient for 3 Quarts*)

2 qt. water 1 Tb. lye

1 qt. shelled corn 3 tsp. salt

Put the water into a large kettle or saucepan, and into the water put the lye. Allow the water to come to the boiling point, and then add the corn and let it boil until the skins will slip off the grains when they are pressed between the thumb and the finger. Take from the stove, stir sufficiently to loosen the skins, and then remove them by washing the grains of corn in a coarse colander. Cover the grains with cold water and return to the fire. When the water boils, pour it off. Repeat this process at least three times, so as to make sure that there is no trace of the lye, and then allow the grains to cook in more water until they burst. Season them with the salt, and while the hominy thus prepared is still hot put it into a jar or a crock and cover it tight until it is to be used. The water in which the hominy is cooked should remain on it.

**Buttered Hominy**

Perhaps the simplest method of preparing cooked hominy is to butter it. In this form it may be served with cream as a breakfast or a luncheon dish, or it may be used in the place of a vegetable.

**Buttered Hominy**

(*Sufficient to Serve Six*)

1 pt. cooked hominy 3 Tb. butter 1 tsp. salt

Allow a few spoonfuls of water to remain on the cooked hominy. Add the butter and the salt, and then heat all thoroughly, stirring the hominy gently so as to incorporate, or mix in, the butter and the salt. Serve while hot.

**Creamed Hominy**

The addition of a cream sauce to cooked hominy not only adds to the palatableness of this cereal, but increases its food value. When hominy is served with a sauce, it may be used as a dinner vegetable or as the main dish in a light meal.

**Creamed Hominy**

(*Sufficient to Serve Six*)

1 c. milk

2 Tb. butter 1 tsp. salt 1 Tb. flour 1 pt. cooked hominy

Heat the milk, and to it add the butter and the salt. Then thicken it with the flour. To this sauce add the hominy and allow all to cook slowly for 10 or 15 minutes. Serve the creamed hominy hot.

HOMINY GRITS.—The cereal sold under the name of *hominy grits* is prepared commercially by crushing dried hominy grains. It has practically the same food value as hominy, and in appearance resembles cream of wheat. The following recipe shows the simplest way in which to prepare this food, it being usually served as a breakfast cereal in this form:

## Hominy Grits

(*Sufficient to Serve Six*)

1 tsp. salt

4 c. water

1 c. hominy grits

Add the salt to the water and bring it to the boiling point. Stir the hominy grits into the water and continue to boil for 10 minutes. Then place in a double boiler and cook for 3 to 4 hours. Serve hot with cream or milk and sugar.

## Left-over Hominy

No waste need result from hominy that is not used at the meal for which it is prepared, for it may be utilized in many ways. For example, it may be served cold with fruit and cream, made into croquettes with chopped meat or cheese and either sautéd or baked, or used in soups to increase materially their food value. A dish prepared by combining cooked or left-over hominy with other ingredients to form hominy and cheese soufflé, which is illustrated in Fig. 2, will prove to be very appetizing.

## Hominy and Cheese Soufflé

(*Sufficient to Serve Six*)

1-1/2 c. cooked hominy 1/2 c. hot milk 1/2 tsp. salt 1/2 tsp. paprika

1 c. grated cheese

2 eggs

Work the hominy smooth by mashing it with a fork, and then add the hot milk, salt, paprika, and grated cheese. Separate the eggs, beat the yolks thoroughly, and stir them well into the mixture. Next, fold in the whites, which should be stiffly beaten, pour the mass into a buttered baking dish, and bake until it is firm in the center. Serve hot.

CORN-MEAL MUSH.—Since corn meal is comparatively inexpensive and high in food value, the housewife can make frequent use of it to advantage. In the form of mush, corn meal is easily digested; besides, such mush is a very good breakfast cereal when served hot with milk or cream. Although the recipe here given makes a sufficient amount for six persons, a good plan is to increase the quantities mentioned so that there will be enough mush left to mold and use in other ways.

**Corn-meal Mush**

(*Sufficient to Serve Six*)

1 tsp. salt 3-1/2 c. water 1 c. corn meal

Add the salt to the water and bring the salted water to the boiling point. When it is boiling rapidly, sift the corn meal slowly through the fingers into it, and at the same time stir it rapidly so as to prevent the formation of lumps. Any mush that contains lumps has not been properly made and should not be served in this condition, as it is unpalatable. Keep stirring constantly until the corn meal thickens; then place it in a double boiler and allow it to cook from 2 to 4 hours, when it should be ready to serve. This method of cooking mush is the most convenient, because not much stirring is required after the corn meal is thickened. A heavy aluminum kettle or an iron pot is a good utensil in which to cook mush, as it does not burn easily in either, although almost constant stirring is required. When the mush becomes very thick, the heated air, in forcing its way through the mush in the process of boiling, makes the mush pop and very often splash on the hands and burn them. To avoid such an accident, therefore, it is advisable to wrap the hand used for stirring in a towel or a cloth.

**Sautéd Corn-meal-mush**

Mush cooked in the manner just explained may be poured into pans, such as bread pans, where it will harden and form a mold that can be sliced as thick or as thin as desired and then sautéd. Corn-meal mush prepared in this way pleases the taste of many persons, and while some persons find it harder to digest than just plain mush, it serves to give variety to meals. For sautéing mush, a heavy iron or steel frying pan or griddle should be used, because utensils made of thin material will allow the mush to burn before it browns properly. Put enough fat, such as lard, cooking oil, or drippings, into the cooking utensil so that when heated it will be about 1/4 inch deep all over the surface. When the utensil is very hot, put in the slices of mush and allow them to brown on one side. Then turn the slices over carefully, so as not to break them, and brown them on the other side. As will be observed, corn-meal mush does not brown quickly in sautéing. This characteristic is due to the large amount of moisture it contains. Serve the mush hot, and to add to its flavor serve with it sirup or honey.

### Corn-meal Croquettes

Croquettes of any kind add variety to a meal, and because they are attractive they appeal to the appetite. To make croquettes of corn meal, mold mush as for sautéing. Then cut this into slices 1 inch thick, and cut each slice into strips 1 inch wide. Roll these in slightly beaten egg and then in crumbs, and sauté them in hot fat until they are crisp and brown. Serve these croquettes hot with either butter or sirup or both.

### Left-over Corn-meal Mush

Sautéd corn-meal mush and corn-meal croquettes can, of course, be made from mush that is left over after it has been cooked to serve as a cereal; however, if there is only a small quantity left, it may be utilized in still another way, namely, as a garnish for the platter on which meat is served. To prepare corn-meal mush in this way, spread it about 1/3 inch thick in a pan and allow it to cool. Then turn it out of the pan in a sheet on a board that has been floured; that is, covered thinly with flour. Cut this sheet of corn meal into small circles with the aid of a round cutter or into diamond shapes with a knife, and then brown both sides of each of these in butter.

## WHEAT

### Origin and Use

WHEAT, owing to the fact that it is grown in all parts of the world and forms the basis for a large amount of the food of most people, is a very important grain. It was probably a native grass of Asia Minor and Egypt, for in these countries it first received cultivation. From the land of its origin, the use of wheat spread over all the world, but it was not introduced into America until after the discovery of this country by Columbus. Now, however, the United States raises more wheat than any other one country, and nearly one-fourth of all that is raised in the world. Wheat is universally used for bread, because it contains a large amount of the kind of protein that lends a rubbery consistency to dough and thus makes possible the incorporation of the gas or air required to make bread light. The use of wheat, however, is by no means restricted to bread, for, as is well known, many cereal foods are prepared from this grain.

In its simplest food form, wheat is prepared by merely removing the coarse bran from the outside of the wheat grain and leaving the grain whole. This is called *hulled,* or *whole, wheat,* and requires soaking or long, slow cooking in order that all its starch granules may be reached and softened sufficiently to make it palatable. The other preparations are made by crushing or grinding the grains from which some of the bran and germ has been removed. Besides flour, which, as has been implied, is not considered as a cereal in the sense used in this Section, these preparations include *wheat grits,* such foods as *cream of wheat* and *farina,* and many *ready-to-eat cereals.* In the preparation of wheat grits, much of the bran is allowed to remain, but neither cream of wheat nor farina contains cellulose in any appreciable quantity. As the addition of bran, however, serves to give these foods bulk, a much more ideal breakfast cereal will result if, before cooking, equal portions of the cereal and the bran are mixed. In preparing ready-to-eat wheat cereals for the market, the manufacturers subject the grains to such elaborate methods of cooking, rolling, and toasting that these foods require but very little additional attention before serving. The only wheat products that demand further attention at this time, therefore, are those which must be cooked before they can be served and eaten.

## RECIPES FOR WHEAT AND WHEAT PRODUCTS

### Hulled Wheat

Inasmuch as hulled, or whole, wheat requires very little preparation for the market, it is a comparatively cheap food. It is used almost exclusively as a breakfast cereal, but serves as a good substitute for hominy or rice. Although, as has been mentioned, it requires long cooking, its preparation for the table is so simple that the cooking need not necessarily increase its cost materially. One of the advantages of this food is that it never becomes so soft that it does not require thorough mastication.

### Hulled Wheat

(*Sufficient to Serve Four*)

1 c. hulled wheat 3 c. water 1 tsp. salt

Look the wheat over carefully and remove any foreign matter. Then add the water and soak 8 to 10 hours, or overnight. Add the salt, cook directly over the flame for 1/2 hour, and then finish cooking in a double boiler for 3 to 4 hours. Serve with cream or milk and sugar.

### Wheat Grits

The cereal known as wheat grits is made commercially by crushing the wheat grains and allowing a considerable proportion of the wheat bran to remain. Grits may be used as a breakfast cereal, when they should be served hot with cream or milk and sugar; they also make an excellent luncheon dish if they are served with either butter or gravy. The fact that this cereal contains bran makes it an excellent one to use in cases where a food with bulk is desired. The accompanying recipe is for a plain cereal; however, an excellent variation may be had by adding 1/2 cupful of well-cleaned raisins 1/2 hour before serving.

### Wheat Grits

(*Sufficient to Serve Four*)

1/2 tsp. salt

3 c. boiling water

3/4 c. wheat grits

Add the salt to the boiling water, sift the wheat grits through the fingers into the rapidly boiling water, and stir rapidly to prevent the formation of lumps. Cook for a few minutes until the grits thicken, and then place in a double boiler and cook 2 to 4 hours.

CREAM OF WHEAT.—In the manufacture of cream of wheat,

not only is all the bran removed, as has been stated, but the wheat is made fine and granular. This wheat preparation, therefore, does not require so much cooking to make it palatable as do some of the other cereals; still, cooking it a comparatively long time tends to improve its flavor. When made according to the following recipe it is a very good breakfast dish:

**Cream of Wheat**

(*Sufficient to Serve Six*)

1 tsp. salt

4-1/2 c. boiling water

3/4 c. cream of wheat

Add the salt to the boiling water, and when it bubbles sift in the cream of wheat through the fingers, stirring rapidly to prevent the formation of lumps. Cook over the flame for a few minutes until it thickens; then place it in a double boiler and cook for 1 to 2 hours. Serve hot with cream or milk and sugar.

**Cream of Wheat With Dates**

Dates added to cream of wheat supply to a great extent the cellulose and mineral salts that are taken out when the bran is removed in the manufacture of this cereal. They likewise give to it a flavor that is very satisfactory, especially when added in the manner here explained.

**Cream of Wheat With Dates**

(*Sufficient to Serve Six*)

3/4 c. cream of wheat 1 tsp. salt

4-1/2 c. boiling water 3/4 c. dates

Cook the cream of wheat in the manner directed in Art. 47. Wash the dates in hot water, cut them lengthwise with a sharp knife, and remove the seeds. Cut each date into four pieces and add them to the cream of wheat 10 minutes before serving, stirring them into the cereal just enough to distribute them evenly. Serve hot with cream or milk and sugar.

**FARINA**

The wheat preparation called farina is very much the same as cream of wheat, being manufactured in practically the same manner. It is a good breakfast cereal when properly cooked, but it does not contain sufficient cellulose to put it in the class of bulky foods. However, as has been pointed out, this bulk may be supplied by mixing with it, before cooking, an equal amount of bran. In such a case, of course, more water will be needed and the cooking process will have to be prolonged. Plain farina should be prepared according to the recipe here given, but, as in preparing cream of wheat, dates may be added to impart flavor if desired.

## Farina

(*Sufficient to Serve Six*)

1 tsp. salt

4 c. boiling water

3/4 c. farina

Add the salt to the boiling water, and as the water bubbles rapidly sift the farina into it slowly through the fingers, stirring rapidly to prevent the formation of lumps. Then place it in a double boiler and allow it cook for 2 to 4 hours. Serve hot with cream or milk and sugar.

## Graham Mush With Dates

Graham flour is a wheat product that is high in food value, because in its manufacture no part of the wheat grain is removed. While the use of this flour as a breakfast cereal is not generally known, it can be made into a very appetizing and nutritious dish, especially if such fruit as dates is mixed with it.

## Graham Mush With Dates

(*Sufficient to Serve Six*)

1-1/4 c. graham flour 3 c. water 1 tsp. salt 1 c. dates

Moisten the graham flour carefully with 1 cupful of the cold water. When perfectly smooth, add it to the remainder of the water, to which the salt has been added, and boil rapidly, allowing the mixture to cook until it thickens. Then place it in a double boiler and cook 1 to 2 hours. Wash the dates, remove the stones, and cut each into four pieces. Add these to the mush 10 minutes before serving. Serve hot with cream or milk and sugar.

## Left-over Wheat Cereals

Numerous ways have been devised for utilizing wheat cereals

that are left over, so that no waste need result from what is not eaten at the meal for which a cereal is cooked. For instance, left-over hulled wheat can be used in soup in the same way as barley and rice, and plain cream of wheat and farina can be molded, sliced, and sautéd like corn-meal mush and served with sirup. The molded cereal can also be cut into 2-inch cubes and served with any fruit juice that is thickened slightly with corn starch. Besides utilizing left-over wheat cereals in the ways mentioned, it is possible to make them into custards and soufflés, as is shown in the two accompanying recipes, in which cream of wheat may be used in the same manner as farina.

## Farina Custard

(*Sufficient to Serve Six*)

1 c. cold farina

2 c. milk 2 eggs 1/2 c. sugar 1/4 tsp. nutmeg

Stir the farina and milk together until they are perfectly smooth; then add the eggs, beaten slightly, the sugar, and the nutmeg. Bake in a moderately hot oven until firm and serve hot or cold with any sauce desired.

## Farina Soufflé

(*Sufficient to Serve Six*)

1 c. cold farina 1-1/2 c. milk 1/2 tsp. salt 1/4 tsp. paprika

1 c. grated cheese

2 eggs

Stir the farina smooth with the milk, add the salt, paprika, grated cheese, and egg yolks, which should first be beaten. Then beat the egg whites stiff and fold them into the mixture. Pour all into a buttered baking dish, place this in a large pan filled with enough hot water to reach almost to the top of the baking dish, and bake in a moderately hot oven until the mixture in the dish is firm in the center. Serve at once upon taking from the oven.

# RICE

## Varieties and Structure

RICE, next to wheat, is used more extensively as a food than any other cereal. It is a plant much like wheat in appearance, but it grows only in warm climates and requires very moist soil. In fact, the best land for rice is that which may be flooded with about 6 inches of water. This cereal is of two kinds, namely, Carolina rice and Japanese rice. *Carolina rice,* which is raised chiefly in the southeastern part of the United States, has a long, narrow grain, whereas *Japanese rice,* which originated in Japan and is raised extensively in that country and China and India, has a short, flat, oval grain. Efforts made to raise the Japanese variety in the United States show a peculiarity of this cereal, for when it is planted in the same locality as Carolina rice, it soon loses its identity and takes on the shape of the other. Although vast crops of rice are raised in the United States, a large quantity of it must be imported, because these crops are not sufficient to supply the demands of this country.

Before rice grains are prepared for use as food, they have two coverings. One is a coarse husk that is thrashed off and leaves the grain in the form of unpolished rice and the other, a thin, brown coating resembling bran. This thin coating, which is very difficult to remove, is called, after its removal, *rice polishings*. At one time, so much was said about the harmful effect of polished rice that a demand for unpolished rice was begun. This feeling of harm, however, was unnecessary, for while polished rice lacks mineral matter to a great extent, it is hot harmful to a person and need cause no uneasiness, unless the other articles of the diet do not supply a sufficient amount of this food substance. After the inner coating has been removed, some of the rice is treated with paraffin or glucose and talc to give it a glazed appearance. This is called *polish,* and is sometimes confounded with the term rice polishings. However, no confusion regarding these terms will result if it is remembered that rice polishings are the thin inner coating that is removed and polish is what is added to the rice. In composition, rice differs from the other cereals in that it is practically all starch and contains almost no fat nor protein.

To be perfect, rice should be unbroken and uniform in size, and in order that it may be put on the market in this form the broken grains are sifted out. These broken grains are sold at a lower price than the whole grains, but the only difference between them is their appearance, the broken grains being quite as nutritious as the whole grains. In either form, rice is a comparatively cheap food, because it is plentiful, easily transported, and keeps perfectly for an indefinite period of time with very little care in storage. Before rice is used, it should be carefully examined and freed from the husks that are apt to remain in it; then it should be washed in hot water. The water in which rice is washed will have a milky appearance, which is due to the coating that is put on in polishing rice.

## Recipes for Rice

Rice may be cooked by three methods, each of which requires a different proportion of water. These methods are *boiling*, which requires twelve times as much water as rice; the *Japanese method*, which requires five times as much; and *steaming*, which requires two and one-half times as much. Whichever of these methods is employed, however, it should be remembered that the rice grains, when properly cooked, must be whole and distinct. To give them this form and prevent the rice from having a pasty appearance, this cereal should not be stirred too much in cooking nor should it be cooked too long.

## Boiled Rice

Boiling is about the simplest way in which to prepare rice for the table. Properly boiled rice not only forms a valuable dish itself, but is an excellent foundation for other dishes that may be served at any meal. The water in which rice is boiled should not be wasted, as it contains much nutritive material. This water may be utilized in the preparation of soups or sauces, or it may even be used to supply the liquid required in the making of yeast bread. The following recipe sets forth clearly how rice should be boiled:

## Boiled Rice

(*Sufficient to Serve Eight*)

1 c. rice

3 tsp. salt

3 qt. boiling water

Wash the rice carefully and add it to the boiling salted water. Boil rapidly until the water begins to appear milky because of the starch coming out of the rice into the water or until a grain can be easily crushed between the fingers. Drain the cooked rice through a colander, and then pour cold water over the rice in the colander, so as to wash out the loose starch and leave each grain distinct.

Reheat the rice by shaking it over the fire, and serve hot with butter, gravy, or cream or milk and sugar.

## Japanese Method Of Cooking Rice

Rice prepared by the Japanese method may be used in the same ways as boiled rice. However, unless some use is to be made of the liquid from boiled rice, the

Japanese method has the advantage of being a more economical way of cooking this cereal.

## Japanese Method

(*Sufficient to Serve Eight*)

1 c. rice 1-1/2 tsp. salt 5 c. boiling water

Wash the rice, add it to the boiling salted water, and boil slowly for 15 minutes. Then cover the utensil in which the rice is cooking and place it in the oven for 15 minutes more, in order to evaporate the water more completely and make the grains soft without being mushy. Serve in the same way as boiled rice.

## Steamed Rice

To steam rice requires more time than either of the preceding cooking methods, but it causes no loss of food material. Then, too, unless the rice is stirred too much while it is steaming, it will have a better appearance than rice cooked by the other methods. As in the case of boiled rice, steamed rice may be used as the foundation for a variety of dishes and may be served in any meal.

## Steamed Rice

(*Sufficient to Serve Six*)

1 c. rice 1-1/2 tsp. salt 2-1/2 c. water

Wash the rice carefully and add it to the boiling salted water. Cook it for 5 minutes and then place it in a double boiler and allow it to cook until it is soft. Keep the cooking utensil covered and do not stir the rice. About 1 hour will be required to cook rice in this way. Serve in the same way as boiled rice.

CREAMED RICE.—To increase the nutritive value of rice, it is sometimes cooked with milk and cream to form what is known as creamed rice. These dairy products added to rice supply protein and fat, food substances in which this cereal is lacking, and also add to its palatability.

## Creamed Rice

(*Sufficient to Serve Six*)

2-1/2 c. milk 1 c. rice 1-1/2 tsp. salt 1/2 c. cream

Heat the milk in the small pan of a double boiler and add to it the rice and salt. Place this pan into the larger one and cook for about 1 hour, or until the rice is soft. Then pour the cream over the rice and cook a few minutes longer. Serve hot.

## Oriental Rice

As rice is a bland food, practically lacking in flavor, any flavoring material that may be added in its preparation or serving aids in making it more appetizing. Oriental rice, which is prepared according to the following recipe, therefore makes a very tasty dish and one that may be used in place of a vegetable for lunch or dinner.

## Oriental Rice

(*Sufficient to Serve Six*)

1 c. rice 2-1/2 c. stock, or meat broth

2 Tb. butter 1 slice onion 1/2 c. canned tomatoes

Steam the rice in the stock until it is soft by the method given for steaming rice. Then brown the butter and onion in a frying pan, add the tomatoes, and heat thoroughly. Pour this mixture into the rice, mix well, and serve.

## Browned Rice

Another way in which to add variety in serving rice is to brown it.

Sufficient browned rice for six persons may be prepared by putting 1 cupful of clean rice in an iron frying pan that contains no fat, placing the pan directly over the flame, and stirring the rice until the grains become an even, light brown.

Rice that has been treated in this way has additional flavor added to it and can be used in the same way as boiled or steamed rice.

SAVORY RICE.—Rice browned in the manner just explained is used in the preparation of savory rice, a dish that serves as a very good substitute for a vegetable. Savory rice may be prepared according to the following recipe:

## Savory Rice

(*Sufficient to Serve Eight*)

1 c. browned rice 2-1/2 c. water

1 tsp. salt 1/2 c. chopped celery

2 Tb. butter 1 small onion, chopped 1/2 c. canned tomatoes 1/4 c. chopped pimiento

Steam the browned rice in the salted water as in steaming rice, and cook the celery, which should be chopped fine, with the rice for the last half hour of the steaming. Brown the butter and add to it the onion finely chopped, the tomatoes, and the pimiento. A few minutes before serving time, add this to the rice, mix well, and serve hot.

## Left-over Rice

There are a variety of ways in which left-over rice may be used. For instance, rice that has been cooked and is not used may be utilized in soups, combined with pancake, muffin, or omelet mixtures, or made into puddings by mixing it with a custard and then baking. It may be served with fruit, made into patties, or combined with tomatoes, cheese, or meat to form an appetizing dish.

As has been shown, rice is one of the cereals that contain very little cellulose. Fruit added to it in the preparation of any dish makes up for this lack of cellulose and at the same time produces a delicious combination. Rice combined with pineapple to form a dish like that shown in Fig. 3 not only is very attractive but meets with the favor of many; besides, it provides a good way in which to utilize left-over rice.

## Rice With Pineapple

(*Sufficient to Serve Six*)

1 c. steamed or creamed rice

1/4 c. sugar

6 rings pineapple

3/4 c. whipped cream

Stir the sugar into the rice and if necessary moisten with a little cream. Shape the rice into six balls of equal size, making them so that they will be about the same in diameter as the rings of the pineapple, and place one in the center of each pineapple ring. Whip the cream with an egg whip or beater until it stands up well, and garnish each dish with the whipped cream before serving.

Another satisfactory dish may be made by combining eggs with left-over rice to form RICE PATTIES. Owing to the protein supplied by the eggs, such a combination as this may be made to take the place of a light meat dish for luncheon or supper, and, to impart additional flavor, it may be served with any sauce desired.

## Rice Patties

(*Sufficient to Serve Six*)

1 c. stale crumbs 1/2 tsp. salt 1/2 tsp. celery salt

2 eggs 2 c. steamed rice

Add 1/2 cupful of the crumbs, the salt, the celery salt, and the eggs, slightly beaten, to the cold steamed rice. If more moisture seems to be necessary, add a very little milk.

Shape the rice with the other ingredients into round patties, and then roll these in the remainder of the crumbs and sauté them in hot butter. Serve the patties hot and with sauce, if desired.

Besides left-over rice, small quantities of one or more kinds of left-over meat and stock or gravy can be used to make a very appetizing dish known as SPANISH RICE, which may be used as the main, or heavy, dish in a luncheon.

## Spanish Rice

(*Sufficient to Serve Six*)

1 small onion

2 Tb. butter 1-1/2 c. steamed or boiled rice

1 c. chopped meat 1/2 c. meat stock or gravy 1/2 c. canned tomatoes

2 Tb. grated cheese 1/4 c. stale crumbs

Chop the onion and brown it in butter. Mix well the browned onion, rice, chopped meat, stock or gravy, and tomatoes, and pour all into a buttered baking dish. Then sprinkle the cheese and crumbs on top of the mixture and bake for 1 hour in a slow oven. Serve hot.

## OATS

### Composition and Varieties

As an article of food, OATS are used very extensively. In Scotland, this cereal formed the principal article of diet for many years, and as the hardiness of the Scotch people is usually attributed to their diet the value of oats as a food cannot be overestimated. This grain, or cereal, grows very much like wheat and yields an abundant crop in fairly good soil; but it is unlike wheat in composition, for it contains very little protein and considerable fat. In fact, it contains more fat than any other cereal. Because of its lack of protein, it will not make raised bread, and when it must serve the purpose of bread it is made into flat cakes and baked. Although it is used to some extent in this way, its greatest use for food, particularly in the United States, is in the form of *oatmeal* and *rolled oats*. In the preparation of oatmeal for the market, the oat grains are crushed or cut into very small pieces, while in the preparation of rolled oats they are crushed flat between large rollers.

### Recipes for Oats

The same methods of cooking can be applied to both oatmeal and rolled oats. Therefore, while the recipes here given are for rolled oats, it will be well to note that they can be used for oatmeal by merely substituting this cereal wherever rolled oats are mentioned.

ROLLED OATS.—Because of the high food value of rolled oats, this cereal is excellent for cold weather, especially when it is served with hot cream or milk and sugar. It can be prepared very easily, as the accompanying recipe shows.

### Rolled Oats

(*Sufficient to Serve Six*)

1 c. rolled oats 3 c. boiling water

1 tsp. salt

Stir the oats into the boiling water to which the salt has been added. Boil 2 minutes, stirring them occasionally to keep them from sticking. Then cook them in a double boiler for 2 to 4 hours. During this time, stir the oats as little as possible, so as to prevent them from becoming mushy. Serve hot.

ROLLED OATS WITH APPLES.—The combination of rolled oats and apples is rather unusual, still it makes a dish that lends variety to a breakfast or a luncheon. Such a dish is easily digested, because the apples supply to it a considerable quantity of cellulose and mineral salts.

## Rolled Oats With Apples

(*Sufficient to Serve Six*)

2/3 c. rolled oats

2 c. boiling water

1/2 tsp. salt

6 medium-sized apples 1 c. water 1/2 c. sugar

Stir the rolled oats into the boiling salted water and cook them until they set; then place them in a double boiler and cook for 2 to 4 hours. Pare and core the apples, and then cook them whole in a sirup made of 1 cupful of water and 1/2 cupful of sugar until they are soft, but not soft enough to fall apart. To serve the food, place it in six cereal dishes. Put a large spoonful of the cooked oats in each dish, arrange an apple on top of the oats, and then fill the hole left by the core with rolled oats. Over each portion, pour some of the sirup left from cooking the apples, and serve hot with cream.

ROLLED-OATS JELLY WITH PRUNES.—If an appetizing dish for warm weather is desired, rolled oats may be cooked to form a jelly and then have stewed prunes added to it. Such a dish is illustrated in Fig. 4. When served with cream, this combination of rolled oats and prunes is high in food value and consequently may be made the important dish in the meal for which it is used.

## Rolled-oats Jelly With Prunes

(*Sufficient to Serve Six*)

1 c. rolled oats

3 c. water

1 tsp. salt

12 stewed prunes

Cook the rolled oats according to the directions already given, and then force them through a fine sieve. Remove the seeds from the prunes that have been stewed by cooking them very slowly until they are soft in a sufficient quantity of water to cover them well, drain off all the juice, and place two prunes in the bottom of each of six cups, or molds, that have been moistened with cold water. Fill each with the rolled-oats jelly and set them aside to chill. When ready to serve, turn the food out of each mold into a cereal dish and serve with cream and sugar.

LEFT-OVER ROLLED OATS.—Every housewife should refrain from throwing away any left-over rolled oats, because all of this cereal remaining from a previous meal can be used to good advantage. For example, it can be made especially tasty if, before it is cold, it is added to fruit, poured into molds and allowed to stand in them until it is cold, and then served with sugar and cream.

Fruits of any kind, such as cooked peaches, prunes, and apricots or fresh bananas, may be used for this purpose by cutting them into small pieces. Another way of utilizing this cereal when it is warm is to pour it into a pan or a dish, press it down until it is about 1 inch thick, and then, after it is cold, cut it into pieces of any desirable size or shape, brown these pieces in butter, and then serve them with sirup. If the left-over cereal is cold, a good plan would be to serve it with baked apple; that is, for each person to be served, place a spoonful of the cereal in a dish with a baked apple, sprinkle a little cinnamon or nutmeg over it, and then serve it with cream. Still another very good way in which to utilize left-over rolled oats is to make it into croquettes according to the following recipe:

## Rolled-oats Croquettes

(*Sufficient to Serve Four*)

1/2 c. grated cheese

3/4 c. crumbs

1/2 tsp. salt

1/4 tsp. paprika

1 c. cooked rolled oats

1 egg

Work the cheese with 1/2 cupful of the crumbs, the salt, and the paprika into the cold rolled oats; then add the egg, which should be slightly beaten. If more moisture seems to be necessary, add a little milk. Form the ingredients into small croquettes, and then roll them in the remaining 1/4 cupful of crumbs and sauté then in butter. Garnish with parsley and serve.

## BARLEY

### Origin and Use

BARLEY is a grain, or cereal, that grows very much like wheat. However, it is hardier than wheat or any other cereals and may be grown through a greater range of climates. Barley has been cultivated from the most ancient times; in fact, its cultivation can be traced as far back as man's occupations have been recorded. The grain of this cereal has also played an important part in the advancement of man, for, according to history, some of the present weights and measures originated from it. Thus, the Troy weight grain is said to have been first fixed by finding the average weight of a barley grain, and the inch of linear measure, by placing three grains of barley end to end.

Although several varieties of barley have been cultivated as food from the earliest times, the grain is now used principally in the manufacture of malt. In this form, it is used for the malting of foods and in the making of alcoholic liquors. To produce malt, the barley grains are moistened and allowed to sprout, and during this process of sprouting the starch of the barley is changed to sugar. The grains are then dried, and the sprouts, which are called *malt sprouts*, are broken off and sold as cattle food. The grain that remains, which is really *malt*, is then crushed and combined with other grains for use as malted cereal food. When barley is used to make malt, or fermented, liquors, it is soaked in water, which absorbs the sugar in it; then yeast is added, and this produces alcohol by causing the fermentation of the sugar.

In the United States, *pearl barley* is the name applied to the most common form of barley used as food. In this form, the layer of bran is removed from the outside of the barley grain, but no change is made in the grain itself. Pearl barley is used for soups and as a breakfast cereal, but for whatever purpose it is employed it requires very long cooking to make it palatable. Very often the water in which a small amount of pearl barley has been cooked for a long time is used to dilute the milk given to a child who has indigestion or who is not able to take whole milk.

## Recipes for Barley

PEARL BARLEY.—As a breakfast cereal, possibly the only satisfactory way in which to prepare pearl barley is to cook it in a double boiler, although after it is cooked in this way it may, of course, be used to prepare other breakfast dishes. Barley is not liked by everybody; nevertheless, it is an excellent food and its nature is such that even after long cooking it remains so firm as to require thorough mastication, which is the first great step in the digestion of starchy foods.

## Pearl Barley

(*Sufficient to Serve Six*)

1 c. pearl barley

1 tsp. salt

4-1/2 c. boiling water

Look the barley over carefully and remove any foreign particles it may contain. Add it to the boiling salted water, and cook it directly over the flame for 10 minutes. Then place it in a double boiler and cook for 3 to 4 hours. For the barley to be cooked properly, the water should be completely absorbed. Serve hot with cream or milk and sugar.

PEARL BARLEY WITH FRUIT.—Cooked barley does not contain very much flavor. Therefore, if a more tasty dish is desired, it is usually necessary to add something, such as fruit, that will improve the flavor. Various fruits may be used with barley, as is shown in the accompanying recipe.

## Pearl Barley With Fruit

(*Sufficient to Serve Eight*)

1 c. pearl barley 1 tsp. salt

5 c. boiling water

1 c. dates, figs, or prunes

Examine the barley to see that it contains no foreign matter, and then put it to cook in the boiling water to which the salt has been added. After cooking directly over the flame for 10 minutes, place it in a double boiler and cook it for 3 to 4 hours. If dates are to be used, wash them in warm water, remove the seeds, and cut each into four pieces. In the case of figs, soak them in hot water for 1/2 hour and then cut them into small pieces. If prunes are desired, stew them as explained in Art. 71, and when the seeds are removed cut them into small pieces. Add the fruit to the barley 10 or 15 minutes before removing it from the stove. Serve hot with cream or milk and sugar.

LEFT-OVER BARLEY.—Cooked barley that is left over from a meal should not be wasted. That which has been cooked without fruit may be added to meat stock or used with vegetables for soup. Also, cooked barley that has had time to set and become stiff may be sautéd in butter until it is slightly brown. When served with meat gravy, barley prepared in this manner makes a very appetizing and satisfying luncheon dish.

## Rye, Buckwheat, and Millet

RYE is a grain that grows very much like wheat, but it can be cultivated in poorer soil and colder climates than this cereal. It is not used alone to any great extent for anything except the making of bread, but it is particularly well adapted for this purpose, since it contains a large amount of gluten, the food substance necessary for successful bread making, and, like wheat, will make yeast bread when used alone. Bread made of rye flour has a dark color and a peculiar flavor, and while these characteristics make it unpopular with some persons it is used extensively by certain classes, especially persons from foreign countries. Besides its use for bread, rye is frequently combined with other cereals in the manufacture of ready-to-eat cereal foods.

BUCKWHEAT is used less extensively than any of the other cereals already mentioned, but it has an advantage over them in that it thrives in soil that is too poor for any other crop. The buckwheat plant grows to a height of about 2 feet and blossoms with a white flower. Its seeds, which are three-cornered in shape, bear a close resemblance to beechnuts, and because of this peculiar similarity, this cereal was originally called *beech wheat*. Practically the only use to which buckwheat is put is to grind it into very fine flour for griddle cakes, recipes for which are given in another Section.

MILLET as a cereal food finds practically no use in the United States; in fact, in this country it is grown almost exclusively for cattle food, the stalk of the plant being large and juicy and containing a considerable amount of food. The seed of this plant furnishes the smallest grain known for use as food, and because of its size it is very hard to gather. Millet, however, is used extensively by some of the people of Southern Asia and India, who depend on it very largely, since, in some localities, it forms their only cereal food. In these countries, it is ground into flour and used for making bread.

## Prepared, Or Ready-to-eat, Cereals

All the cereals that have been discussed up to this point require cooking; but there are many varieties of cereal food on the market that are ready to eat and therefore need no further preparation. Chief among these are the cereal foods known as *flakes*. These are first made by cooking the grain, then rolling it between rollers, and finally toasting it. The grains that are treated in this way for the preparation of flake foods are wheat, corn, rye, and rice. It is well to remember this fact, because the trade name does not always indicate the kind of grain that has been used to make the food. In another form in which cereals, principally wheat, appear on the market, they are cooked, shredded, pressed into biscuits, and then toasted. Again, cereals are made into loaves with the use of yeast, like bread, and after being thoroughly baked, are ground into small pieces. Wheat generally forms the basis of these preparations, and to it are added such other grains as rye and barley.

The toasting of cereals improves their flavor very materially and at the same time increases their digestibility. In fact, cereals that have been subjected to this process are said to be predigested, because the starch granules that have been browned in the toasting are changed into *dextrine,* and this is one of the stages through which they must pass in their process of digestion in the body. However, the housewife should not allow herself to be influenced unduly by what is said about all prepared cereals, because the manufacturer, who has depended largely on advertising for the sale of his product, sometimes becomes slightly overzealous and makes statements that will bear questioning. For instance, some of these foods are claimed to be muscle builders, but every one should remember that, with the exception of rye and wheat, which build up the tissues to a certain extent, the cereals strengthen the muscles in only a slight degree. Others of these foods are said to be nerve and brain foods, but it should be borne in mind that no food acts directly on the nerves or the brain. In reality, only those foods which keep the body mentally and physically in good condition have an effect on the nerves and the brain, and this at best is an indirect effect.

## Serving Cereals

Although, as is shown by the recipes that have been given, cereals may have a place in practically all meals that the housewife is called on to prepare, they are used more frequently for breakfast than for any other meal. When a cereal forms a part of this meal, it should, as a rule, be served immediately after the fruit, provided the breakfast is served in courses. Many persons, of course, like fresh fruit served with cooked or dry cereal, and, in such an event, the fruit and cereal courses should be combined. A banana sliced over flakes or a few spoonfuls of berries or sliced peaches placed on top afford a pleasing change from the usual method of serving cereals. Another way in which to lend variety to the cereal and at the same time add nourishment to the diet is to serve a poached egg on top of the shredded-wheat biscuit or in a nest of corn flakes, especially if they have been previously heated. In fact, any of the dry cereals become more appetizing if they are heated thoroughly in a slow oven and then allowed to cool, as this process freshens them by driving off the moisture that they absorb and that makes them tough.

To add to both dry and cooked cereals protein and fat, or the food elements in which they are not so high, milk or cream is usually served with them. Of these dairy products, which may be served hot or cold, milk adds more protein than cream, and cream more fat than milk. Some persons, however, who do not care for milk and cream or cannot take them, substitute a little butter for them or find fruit juice a very good accompaniment, especially to a dry cereal. Sugar is generally served with both kinds of cereals, as the majority of persons prefer them slightly sweet; but there is no logical reason for its use except to add flavor.

# 5

# Bread

## BREAD-MAKING REQUIREMENTS

### Importance of Bread as Food

BREAD is sometimes defined as any form of baked flour, but as the word is commonly understood it means only those forms of baked flour which contain some leavening substance that produces fermentation. The making of bread has come down through the ages from the simplest methods practiced by the most primitive peoples to the more elaborate processes of the present day. In truth, to study the history of bread making would amount to studying the accounts of the progress that has been made by the human race. Still, in order that the production of bread from suitable ingredients may be fully understood, it will be well to note the advancement that has been made.

In the earliest times, what was used as bread was made in much the same way as it is today by many uncivilized and semicivilized people. The grain was ground between stones, usually by hand, and then mixed with water to form a dough; then this dough was formed into flat, compact cakes and baked in hot ashes, the result being a food very difficult to digest. Later on, some one discovered that by allowing the dough to stand until fermentation took place and then mixing it with new dough, the whole mass would rise, and also that by subjecting this mass to the action of heat, that is, baking it, the mass would be held in place and become a loaf of raised bread that was lighter and, of course, more digestible. It was this discovery that led up to the modern bread-making processes, in which substances known as *leavening agents*, or *ferments*, are used to make bread light, or porous. Chief among the substances is yeast, a microscopic plant that produces fermentation under favorable conditions.

Indeed, so important is this ferment that, in the United States, whenever the term *bread* is used alone it means *yeast*, or *leavened, bread,* whereas, when other leavening agents are used, the bread is referred to as *hot bread,* or *quick bread,* as is fully explained in another Section. It will be well to note this fact, for in all cases throughout these cookery lessons yeast, or leavened, bread is always meant when the term bread is used alone.

References in the history of the ancient Hebrews show that bread made light by means of fermentation was known thousands of years ago, but it was not until after the accidental discovery of the action of yeast that the making of wholesome and digestible bread became possible. Through this important advance in the making of bread came a demand for better grains and more improved methods of making flour. Indeed, so much attention has been given to these matters that at present the three important processes relating to bread-making—the raising of wheat, the milling of flour, and the manufacture of yeast—are carefully and scientifically performed. These industries, together with the commercial manufacture of bread, occupy an important place in the business of practically all civilized nations.

Among people who are not highly civilized, bread forms the chief article of food and often almost the entire diet, even at the present time; but as man progresses in civilization he seems to require a greater variety of food, and he accordingly devises means of getting it. Since bread is only one of the many foods he finds at his disposal, it does not assume a place of so much importance in present-day meals as it formerly did. However, it still makes up a sufficient proportion of the food of every family to warrant such careful and extensive study, as well as such mastery of the processes involved, that the housewife may present to her family only the best quality of this food.

Although it does not have such extensive use as it had in the past, bread of some description, whether in the form of loaves, biscuits, or rolls, forms a part of each meal in every household. This fact proves that, with the exception of milk, it is more frequently eaten than any other food. A food so constantly used contributes very largely to the family's health if it is properly made. However, there is possibly nothing in the whole range of domestic life that so disturbs the welfare of the entire family as an inferior quality of this food, which, besides proving detrimental to the digestion, adds materially to the household expense.

Of course, in many bakeries, bread of an excellent quality is made in a perfectly hygienic manner, and to be able to procure such bread is a wonderful help to the busy housewife or to the woman who finds it inconvenient to make her own bread. Still, practically every person enjoys "home-made" bread so much more than what is made commercially that the housewife will do well to make a careful study of this branch of cookery. If it is properly understood, it will not be found difficult; but the woman who takes it up must manifest her interest to master a few essential principles and to follow them explicitly. After she has obtained the knowledge that she must possess, experience and practice will give her the skill necessary to prevent poor results and a consequent waste of material.

## INGREDIENTS FOR BREAD MAKING

### Ingredients Required

Possibly the first essential to a correct knowledge of bread making is familiarity with the ingredients required. These are few in number, being merely flour, liquid, which may be either milk or water, sugar, salt, and yeast; but the nature of these, particularly the flour and the yeast, is such as to demand careful consideration. It will be admitted that the more the housewife knows about *bread-making materials and processes the greater will be her success in this work. Likewise, it is extremely important that this food be made just as wholesome as possible, for next to milk and eggs, bread ranks as a perfect food, containing all the elements necessary for the growth of the body. This does not mean, though, that any of these foods used as the sole article of diet would be ideal, but that each one of them is of such composition that it alone would sustain life for a long period of time.

### Flour

GRAINS USED FOR FLOUR.—As has been pointed out elsewhere, numerous grains are raised by man, but only two of them, namely, wheat and rye, are used alone for the making of yeast, or leavened, bread. The other grains, such as corn, rice, and oats, produce a flat, unleavened cake, so they are seldom used for bread making unless they are mixed with white flour. Wheat and rye have been used for bread making for a very long time, and their universal use today is due to the fact that they contain considerable protein in the form of *gluten*. This is the substance that produces elasticity in the dough mixture, a condition that is absolutely essential in the making of raised bread. In fact, the toughness and elasticity of bread dough are what make it possible for the dough to catch and hold air and gas and thus produce a light, porous loaf.

Of these two grains, rye is used less extensively in the United States for the making of bread than wheat, although in some countries, particularly the inland countries of Continental Europe, considerable use is made of it. Its limited use here is undoubtedly due to the fact that when rye is used alone it makes a moist, sticky bread, which is considered undesirable by most persons. The reason for this is that, although rye contains a sufficient quantity of gluten, this substance is not of the proper quality to make the elastic dough that produces a light, spongy loaf. Therefore, when rye is used, wheat flour is generally mixed with it. The result is a bread having a good texture, but the dark color and the typical flavor that rye produces.

Wheat, the other grain used for bread making, is an annual grass of unknown origin. It is used more extensively for food than any other grain. In fact, it has been estimated that the average quantity consumed by each person is about 6 bushels a year, and of this amount by far the greater part is used in the making of bread. Since so much of this grain is used as food, considerable time and effort have been spent in developing those qualities which are most desirable for the purpose to which wheat is put and in perfecting the processes whereby wheat flour of a good quality may be obtained.

This grain is particularly well adapted for bread making because of the nature of the proteins it contains and the relative proportions of these. These proteins, which occur in the wheat grain in the form of gluten, are known as *gliadin* and *glutenin*. The gliadin imparts elasticity and tenacity, or toughness, to the gluten, and the glutenin gives it strength. It is not, however, so much the quantity of gluten in the wheat grain that actually determines the quality of flour as the fact that the two varieties must be present in the proper proportions in order for the gluten to have the properties desired for bread making.

Wheat consists of numerous varieties, but only two of these are grown and used in the United States, namely, *spring*, or *hard, wheat* and *winter*, or *soft, wheat*.

SPRING, OR HARD WHEAT is so named because it is sown in the spring of the year and is very tough or firm. Before this variety was known, the wheat used for bread making was not ideal, and the efforts that were made to produce a grain that would be suitable for this purpose resulted in this variety. To obtain its particular composition, spring wheat must be grown under suitable climatic and soil conditions. In North America, it grows in the north central part of the United States and along the southern border of Canada. This variety, which is harvested in the late summer, is characterized by a large proportion of gluten and a correspondingly small amount of starch. It is the presence of the gluten that accounts for the hardness of the spring-wheat grain and the tough, elastic quality of the dough made from the spring-wheat flour. Bread dough, to be right, must have this quality, so that the flour made from spring wheat is used almost exclusively for bread; whereas, for cake and pastry, which should have a tender, unelastic texture, flour made from soft wheat is more satisfactory.

WINTER, OR SOFT WHEAT derives its name from the fact that it is planted in the autumn and is soft in texture. It is of less importance in the making of bread than spring, or hard, wheat, but it is the kind that has been grown for centuries and from which the varieties of spring wheat have been cultivated. It is a softer grain than spring wheat, because it contains less gluten and more starch. The flour made from it does not produce so elastic a dough mixture as does that made from the other variety of wheat; consequently, the finished product, such as bread, rolls, etc., is likely to be more tender and more friable, or crumbly. It is for this reason that winter, or soft, wheat is not used extensively for bread, but is employed for pastry flour or mixed with spring wheat to make what is called a *blend flour*, which may be used for all purposes.

STRUCTURE OF WHEAT GRAIN.—In its natural state, wheat contains all the food substances required for the nourishment of the human body in nearly the proper proportions, and in addition it has in its composition sufficient cellulose to give it considerable bulk. It has been estimated that the average composition of this grain is as follows:

| | PER CENT. |
|---|---|
| Protein | 11.9 |
| Fat | 2.1 |
| Carbohydrates | 71.9 |
| Mineral salts | 1.8 |
| Water | .5 |
| Cellulose | 1.8 |
| Total | 100.0 |

So that the composition of wheat and the making of wheat flour may be more clearly understood, it will be well to observe the structure of a grain, or kernel, of wheat, which is shown greatly enlarged in Fig. 1. At *a* is shown the germ of the young plant, which remains undeveloped until the grain is planted. This part contains practically all the fat found in the grain, some starch, and a small quantity of protein. At *b* is shown the inside of the kernel, or the *endosperm*, as it is called, which is composed of starch granules interlaced with protein and mineral salts. Surrounding these, as at *c*, is a layer of coarse cells that contain mineral matter and protein, and between these cells and the outer husk, as at *d, e, f,* and *g,* are layers of bran, which are composed of cellulose and contain mineral salts and small quantities of starch and protein. Enveloping the entire kernel is a husk, or bran covering, *h*. This forms a protection to the rest of the grain, but it cannot be used as food, because it is composed almost entirely of cellulose, which is practically indigestible. The center of the grain, or the heart, is the softest part and consists of cells filled with starch. From this soft center the contents of the grain gradually grow harder toward the outside, the harder part and that containing the most gluten occurring next to the bran covering.

MILLING OF WHEAT FLOUR.—Great advances have been made in the production of flour from wheat, and these are very good evidence of man's progress in the way of invention. The earliest method consisted in crushing the grain by hand between two stones, and from this crude device came the mortar and pestle. A little later millstones in the form of thick, heavy disks were brought into use for grinding grain. Two of these stones were placed so that their surfaces came together, the lower one being stationary and the upper one made to revolve. Early grinding apparatus of this kind was turned by human power, but this kind of power was first displaced by domestic animals and later by wind and water. Out of this arrangement, which is still used to some extent in small mills, has grown the present-day complicated machinery of the roller process, by which any part of the grain may be included or rejected.

In the roller process, the grain is crushed between metal rolls instead of being ground between stones. It is first screened in order to separate all foreign matter from it, and then stored in bins. When it is taken from these receptacles, it is put through another cleaning process, called *scouring*, or it is thoroughly washed and dried in order to loosen the dirt that clings to it and to free it entirely from dust, lint, etc.

As soon as it is completely cleansed, it is softened by heat and moisture and then passed through a set of corrugated rollers, which are adjustable as are the rubber rollers of a clothes wringer and which flatten and break the grains. After this first crushing, some of the bran is sifted out, while the main portion of the grain is put through another set of rollers and crushed more finely. During the milling, these processes of crushing the grain and removing the bran are repeated from six to nine times, each pair of rollers being set somewhat closer than the pair before, until the grain is pulverized. After the grain has been thus reduced to a powder, it is passed through bolting cloth, which acts as a very fine sieve and separates from it any foreign material that may remain. The result is a very fine, white flour.

GRAHAM FLOUR.—Sometimes the entire grain, including the bran, germ, etc., is ground fine enough merely for baking purposes and is used as flour in this form. Such flour is called graham flour. It contains all the nutriment, mineral matter, and cellulose of the original grain, and is therefore considered valuable as food. However, the objection to this kind of flour is that its keeping quality is not so good as that of the kinds from which the germ has been removed, because the fat contained in the germ is liable to become rancid.

WHOLE-WHEAT FLOUR.—The best grades of fine white flour make bread of excellent quality, but such bread is not so nutritious as that made from whole-wheat flour. In the making of this kind of flour, some of the choicest varieties of wheat are first moistened in order to soften the woody fiber of the bran and are then sifted until the outer husk of the grain is removed. After this treatment, the grains are dried and then pulverized into various grades of so-called whole-wheat flour.

The name whole-wheat flour is misleading, because it implies that all of the grain is used; whereas, since several of the outer layers of bran and the germ are removed in its production, whole-wheat flour is merely flour in which practically all the gluten and the starch are retained. Because this variety is not sifted as are the white flours, it is not so fine as they are; but it is not so coarse as graham flour, nor is bread made from it so dark in color. Both graham and whole-wheat flours produce a more wholesome bread than any of the varieties of white flour, because they contain more of the nutritive elements and mineral salts, which are necessary in the diet. The bran that is retained in them is not used by the body as food, but it adds bulk to the diet and assists in carrying on the normal functions of the digestive tract.

SELECTION OF FLOUR.—If a large quantity of flour must be bought at one time, as, for instance, enough to last through an entire season, it is advisable to test it carefully before the purchase is made, so as to avoid the danger of getting a poor grade. As a rule, however, housewives are obliged to purchase only a small quantity at a time. In such cases, it will not be necessary to test the flour before purchasing it, provided a standard make is selected. Very often, too, a housewife in a small family finds it inconvenient to keep on hand a supply of both bread flour and pastry flour. In such an event, a blend flour, which, as has been mentioned, is a mixture of flour made from spring and winter wheat that will do for all purposes, is the kind to purchase. While such flour is not ideal for either bread or pastry, it serves the purpose of both very well.

QUALITY OF FLOUR.—Flour is put on the market in various grades, and is named according to its quality. The highest grade, or best quality, is called *high-grade patent;* the next grade, *bakers';* and the next, *second-grade patent*. The lowest grade, or poorest quality, is called *red dog*. This grade is seldom sold for food purposes, but it is used considerably for the making of paste.

The quality of flour used in bread making is of very great importance, because flour of poor quality will not, of course, make good bread. Every housewife should therefore be familiar with the characteristics of good flour and should buy accordingly.

Several tests can be applied to flour to determine its kind and its quality. The first test is its color. Bread flour, or flour made from spring wheat, is usually of a creamy-white color, while pastry flour, or that made from winter wheat, is more nearly pure white in color. A dark, chalky-white, or gray color indicates that the flour is poor in quality. The second test is the feel of the flour. A pinch of good bread flour, when rubbed lightly between the thumb and the index finger, will be found to be rather coarse and the particles will feel sharp and gritty. When good pastry flour is treated in the same way, it will feel smooth and powdery. The third test is its adhering power. When squeezed tightly in the hand, good bread flour holds together in a mass and retains slightly the impression of the fingers; poor bread flour treated in the same way either does not retain its shape or, provided it contains too much moisture, is liable to make a damp, hard lump. The odor of flour might also be considered a test. Flour must not have a musty odor nor any other odor foreign to the normal, rather nutty flavor that is characteristic of flour.

The bleaching and adulteration of flour are governed by the United States laws. Bleaching is permitted only when it does not reduce the quality or strength nor conceal any damage or inferiority. Such flour must be plainly labeled to show that it has been bleached.

CARE OF FLOUR.—There is considerable economy in buying flour in large quantities, but unless an adequate storing place can be secured, it is advisable to buy only small amounts at a time. Flour absorbs odors very readily, so that when it is not bought in barrels it should if possible be purchased in moisture-proof bags. Then, after it is purchased, it should be kept where it will remain dry and will not be accessible to odors, for unless the storage conditions are favorable, it will soon acquire an offensive odor and become unfit for use. Flour sometimes becomes infested with weevils, or beetles, whose presence can be detected by little webs. To prevent the entrance of insects and vermin of all kinds, flour should be kept in tightly closed bins after it is taken from the barrels or sacks in which it is purchased. If newly purchased flour is found to be contaminated with such insects, it should be returned to the dealer.

## YEAST

NATURE AND ACTION OF YEAST.—How yeast came to be discovered is not definitely known, but its discovery is believed to have been purely accidental. Some mixture of flour and liquid was probably allowed to remain exposed to the air until it fermented and then when baked was found to be light and porous. Whatever the origin of this discovery was, it is certain that yeast was used hundreds of years ago and that its action was not at that time understood. Even at the present time everything concerning the action of yeast is not known; still continued study and observation have brought to light enough information to show that yeast is the agency that, under favorable conditions, produces light, spongy bread out of a flour mixture.

It has been determined that yeast is a microscopic plant existing everywhere in the air and in dust; consequently, it is found on all things that are exposed to air or dust. In order that it may grow, this plant requires the three things necessary for the growth of any plant, namely, food, moisture, and warmth. Carbohydrate in the form of sugar proves to be an ideal food for yeast, and 70 to 90 degrees Fahrenheit is the temperature at which the most rapid growth occurs. When these conditions exist and a sufficient amount of moisture is provided, yeast grows very rapidly and produces fermentation.

The changes that take place when yeast causes fermentation can be detected very readily by observing the fermenting of fruit juice. As every housewife knows, the first indication of a ferment in fruit juice is the appearance of tiny bubbles, which collect on the sides and the bottom of the vessel containing the fruit and then gradually rise to the top. These bubbles are a form of gas called *carbon-dioxide, or carbonic-acid, gas*. If, after they appear, the juice is tasted, it will be found to be slightly alcoholic and to have a somewhat sour or acid taste. The gas, the acid, and the alcohol thus produced are the three results of the action of the ferment.

When yeast is used in the making of bread out of wheat flour, the changes just mentioned take place. To understand the action of this plant, it will be necessary to remember that wheat contains a large proportion of starch. This substance, however, cannot be acted on by the yeast plant; it must first be changed into sugar. The yeast that is added to the flour changes some of the starch into sugar and transforms the sugar into alcohol and carbonic-acid gas. This gas, which is lighter than the dough, rises, and in its efforts to escape expands the elastic, glutinous dough into a mass of bubbles with thin walls until the dough is two or three times its original bulk. The yeast plants, though, must be well distributed throughout the dough; otherwise, there are likely to be no bubbles in some places and large bubbles with thick walls in others. The gas thus formed is prevented from escaping by the toughness or the elasticity of the gluten, and the spaces that it leaves are what produce a light, porous loaf. When the expansion has gone on long enough, the formation of gas is checked and the ferment is killed by baking the dough in a hot oven. During the baking, the alcohol is driven off by heat, some of the starch is browned and forms the crust, and so little acid is produced in the short time in which the yeast is active that it is not noticeable.

COMMERCIAL YEAST.—When yeast plants are deprived of water and food, they cease to multiply. However, under these conditions, they may be kept alive so that when water and food are again provided they will increase in number and carry on their work. Advantage has been taken of these characteristics of yeast, for although at one time the making of yeast was entirely a household process, it has now, like butter, cheese, canned fruit, etc., become a commercial product. The first yeast put on the market was collected from the surface of the contents of brewers' vats, where it floated in large quantities; but as this was an impure, unreliable product composed of various kinds of bacteria, it is no longer used for the purpose of making bread.

At present, yeast is carefully grown as a pure yeast culture, or product. It is marketed in such a way that when proper food, such as soft dough, or sponge, and a favorable temperature are provided, the plants will multiply and act on the carbohydrate that they find in the food. In fact, the purpose of the well-known process of "setting" a sponge is to obtain a large number of yeast plants from a few.

Commercial yeast is placed on the market in two forms—*moist* and *dry*. Each of these yeasts has its advantages, so that the one to select depends on the method preferred for the making of bread as well as the time that may be devoted to the preparation of this food.

Moist yeast, which is usually called *compressed yeast*, consists of the pure yeast culture, or growth, mixed with starch to make a sort of dough and then compressed into small cakes, the form in which it is sold. The moist condition of this kind of commercial yeast keeps the plants in an active state and permits of very rapid growth in a dough mixture. Consequently, it proves very useful for the rapid methods of making bread. It is soft, yet brittle, is of a grayish-white color, and has no odor except that of yeast.

Since the plants of compressed yeast require very little moisture to make them grow, an unfavorable, or low, temperature is needed to keep the yeast from spoiling; in fact, it is not guaranteed to remain good longer than a few days, and then only if it is kept at a temperature low enough to prevent the plants from growing. This fact makes it inadvisable to purchase compressed yeast at great distances from the source of supply, although it may be obtained by parcel post from manufacturers or dealers.

Dry yeast, the other form of commercial yeast, is made in much the same way as moist yeast, but, instead of being mixed with a small amount of starch, the yeast culture is combined with a large quantity of starch or meal and then dried. The process of drying kills off some of the plants and renders the remainder inactive; because of this, the yeast requires no special care and will keep for an indefinite period of time, facts that account for its extensive use by housewives who are not within easy reach of the markets. However, because of the inactivity of the yeast plants, much longer time is required to produce fermentation in a bread mixture containing dry yeast than in one in which moist yeast is used. Consequently, the long processes of bread making are brought about by the use of dry yeast. If moist yeast is used for these processes, a smaller quantity is required.

LIQUID YEAST.—Some housewives are so situated that they find it difficult to obtain commercial yeast in either of its forms; but this disadvantage need not deprive them of the means of making good home-made bread, for they can prepare a very satisfactory liquid yeast themselves. To make such yeast, flour, water, and a small quantity of sugar are stirred together, and the mixture is then allowed to remain at ordinary room temperature, or 70 degrees Fahrenheit, until it is filled with bubbles. If hops are available, a few of them may be added. When such yeast is added to a sponge mixture, it will lighten the whole amount. Before the sponge is made stiff with flour, however, a little of it should be taken out, put in a covered dish, and set away in a cool, dark place for the next baking. If properly looked after in the manner explained, this yeast may be kept for about 2 weeks.

More certain results and a better flavor are insured in the use of liquid yeast if it is started with commercial yeast, so that whenever this can be obtained it should be used. Then, as just explained, some of the liquid containing the yeast or some of the sponge made with it may be retained for the next baking.

QUALITY OF YEAST.—Of equal importance with the quality of flour is the quality of yeast used in the baking of bread. Yeast is, of course, accountable for the lightness or sponginess of bread, but, in addition, it improves the flavor of the bread if it is of good quality or detracts from the flavor if it is of poor quality. Since the condition of yeast cannot be determined until its effect on the finished product is noted, the housewife should take no chances, but should employ only yeast, whether she uses commercial or liquid, that she knows to be good and reliable. Compressed yeast may be easily judged as to quality. It should be grayish white in color, without streaks or spots, and it should have no sour nor disagreeable odor. If home-made yeast is used and the results obtained are not satisfactory, it may be taken for granted that a fresh supply should be prepared.

## YEAST AIDS

As has already been explained, yeast, in order to grow, requires something on which to feed, and the food that produces the most rapid growth is that which contains carbohydrate. Certain of the carbohydrates, however, prove to be better food and produce more rapid growth than others, and these, which are known as yeast aids, are usually added as ingredients in the making of bread. The ones that are most commonly used are sugar and potato water. Sugar is almost always added, but it should be limited in quantity, because a dough mixture that is made heavy with sugar will rise very slowly. Potato water has been found to be a very satisfactory aid, because the starch of the potato is utilized readily by the yeast. If this aid is to be used, the water in which potatoes are boiled may be saved and, when the ingredients required for the making of bread are mixed, it may be added as a part or all of the liquid required. If it is desired to increase the amount of starch in the potato water, a boiled potato or two may be mashed and added to it.

## MILK AND FAT IN BREAD

Milk is sometimes used as a part or as all of the liquid in bread. While it adds nutritive value and is thought by many persons to improve the texture, it is not absolutely essential to successful bread making. Whenever milk is used, it should first be scalded thoroughly. A point that should not be overlooked in connection with the use of milk is that the crust of milk bread browns more readily and has a more uniform color than that of bread in which water is used as liquid.

Like milk, fat adds nutritive value to bread, but it is not an essential ingredient. If it is included, care should be taken not to use too much, for an excessive amount will retard the growth of the yeast. Almost any kind of fat, such as butter, lard or other clear tasteless fats, or any mixture of these, may be used for this purpose, provided it does not impart an unpleasant flavor to the bread.

## PROPORTION OF BREAD-MAKING MATERIALS

No definite rule can be given for the exact proportion of liquid and flour to be used in bread making, because some kinds of flour absorb much more liquid than others. It has been determined, however, that 3 cupfuls of flour is generally needed for each small loaf of bread. With this known, the quantity of flour can be determined by the amount of bread that is to be made. The quantity of liquid required depends on the quantity and kind of flour selected, but usually there should be about one-third as much liquid as flour.

The particular method that is selected for the making of bread, as is explained later, determines the amount of yeast to be used. If it is desired not to have the bread rise quickly, a small quantity, about one eighth cake of compressed yeast or 2 tablespoonfuls of liquid yeast, is sufficient for each loaf; but if rapid rising is wanted, two, three, or four times as much yeast must be used to produce a sufficient amount of carbon dioxide in less time. It should be remembered that the more yeast used, the more quickly will the necessary gas be created, and that, as has already been shown, it is the formation of gas that makes bread light and porous. In addition to flour, liquid, and yeast, 1 teaspoonful of salt, 1 tablespoonful of sugar, and 1 tablespoonful of fat are the ingredients generally used for each loaf of bread.

## UTENSILS FOR BREAD MAKING

NECESSARY EQUIPMENT.—Not many utensils are required for bread making, but the ones that are needed must be of the right kind if the best results are to be obtained. The necessary equipment is illustrated in Fig. 2. It includes a mixing bowl and cover *a*; a flour sieve *b*; measuring cups *c* of standard size, one for moist and one for dry ingredients, measuring spoons *d*, and a case knife or a spatula *e* for measuring; a long-handled spoon *f* for mixing; and baking, or bread, pans *g*. Unless the table is such that it can be used as a molding board, it will be necessary to provide in addition to the equipment mentioned, a molding board of suitable size.

The mixing bowl may be an earthen one or a metal one like that shown in the illustration. The size of the pans used and the material of which the pans are made should also receive attention. The loaves will be found to bake more quickly and thoroughly if they are not made too large and each one is baked in a separate pan. Pans that are 8 inches long, 3 1/2 inches wide, and 3 inches deep are of a convenient size. They may be made of tin, sheet iron, aluminum, or heat-resisting glass, the only requirements being that all the pans used at one baking be of the same material, because, as heat penetrates some materials more quickly than others, the baking will then be more uniform.

CONVENIENT EQUIPMENT.—While the utensils shown in Fig. 2 are all that are actually required in the making of bread, a bread mixer, one style of which is described in *Essentials of Cookery*, Part 2, will be found extremely convenient by the housewife who must bake large quantities of bread at one time and who has not a great deal of time to devote to the work. This labor-saving device can be used and, of course, often is used by the housewife who makes only a small quantity of bread, as, for instance, two to four loaves; but it is not actually needed by her, as she can handle such an amount easily and quickly.

A *cooler*, which consists of a framework covered with wire

netting and supported by short legs, is also a convenient utensil, as it serves as a good place on which to put baked bread to cool. If one of these devices is not available, however, a substitute can be easily made by stretching a wire netting over a wooden frame.

## BREAD-MAKING PROCESSES

### Acquiring Skill in Bread Making

The nature and the quality of the ingredients required to make bread, as well as the utensils that are needed for this purpose, being understood, it is next in order to take up the actual work of making bread. Several processes are included in this work; namely, making the dough, caring for the rising dough, kneading the dough, shaping the dough into loaves, baking the loaves, and caring for the bread after it is baked. When the finished product is obtained, the loaves are ready to be scored and served. A knowledge of how to carry out these processes is of the utmost importance, for much of the success achieved in bread making depends on the proper handling of the ingredients. Of course, skill in manipulation is acquired only by constant practice, so that the more opportunity the housewife has to apply her knowledge of the processes, the more proficient will she become in this phase of cookery. Each one of the processes mentioned is here discussed in the order in which it comes in the actual work of bread making, and while the proper consideration should be given to every one of them, it will be well, before entering into them, to observe the qualities that characterize good wheat bread.

Good wheat bread may be described in various ways, but, as has been learned by experience and as is pointed out by United States government authorities, probably the best way in which to think of it, so far as its structure is concerned, is as a mass of tiny bubbles made of flour and water, having very thin walls and fixed in shape by means of heat. The size of the cells and the nature of the bubble walls are points that should not be overlooked.

Each loaf should be light in weight, considering its size, should be regular in form, and should have an unbroken, golden-brown crust. The top crust should be smooth and should have a luster, which is usually spoken of as the "bloom" of the crust. Taken as a whole, the loaf should have a certain sponginess, which is known as its elasticity, and which is evidenced by the way in which the loaf acts when it is pressed slightly out of shape. As soon as the pressure is removed, the loaf should resume its original shape. This test should produce the same results when it is applied to small pieces of the crust and to the cut surface of the loaf.

The internal appearance must also receive consideration. To be right, wheat bread should be creamy white in color and should have a definite "sheen," which can best be seen by looking across a slice, rather than directly down into it. As already explained, the holes in it should be small and evenly distributed and their walls should be very thin. These points can be readily determined by holding a very thin slice up to the light.

The flavor of bread is also a very important factor, but it is somewhat difficult to describe just the exact flavor that bread should have in order to be considered good. Probably the best way in which to explain this is to say that its flavor should be that which is brought about by treating the wheat with salt. While such a flavor may not be known to all, it is familiar to those who have tasted the wheat kernel.

## MAKING THE DOUGH

### Preliminary Treatment of Ingredients

The first step in bread making, and without doubt the most important one, is the making of the dough. It consists in moistening the flour by means of a liquid of some kind in order to soften the gluten and the starch, to dissolve the sugar, and to cement all the particles together, and then combining these ingredients. Before the ingredients are combined, however, particularly the flour, the liquid, and the yeast, they must generally be warmed in order to shorten the length of time necessary for the yeast to start growing. Much care should be exercised in heating these materials, for good results will not be obtained unless they are brought to the proper temperature. The flour should feel warm and the liquid, whether it be water or milk, should, when it is added, be of such a temperature that it also will feel warm to the fingers. If water is used, it ought to be just as pure as possible, but if milk is preferred it should be used only after it has been scalded. The yeast should be dissolved in a small quantity of lukewarm water. Hot water used for this purpose is liable to kill the yeast and prevent the bread from rising, whereas cold water will retard the growth of the yeast.

## Combining the Ingredients

As soon as the bread ingredients have received the proper treatment, they are ready to be combined. Combining may be done by two different methods, one of which is known as the *short process* and the other as the *long process*. As their names indicate, these methods are characterized by the length of time required for the bread to rise. Each method has its advantages, and the one to select depends on the amount of time and energy the housewife can afford to give to this part of her work. Persons who use the long process believe that bread made by it tastes better and keeps longer than that made by the short process; whereas, those who favor the short process find that it saves time and labor and are convinced that the quality of the bread is not impaired. The more rapid methods of making breads are possible only when yeast in the active state is used and when more of it than would be necessary in the long process, in which time must be allowed for its growth, is employed. However, regardless of the method followed, all bread mixtures must be begun in the same manner.

LONG PROCESS.—By the long process, there are two ways of combining the ingredients in order to make bread. One is known as the *sponge method* and the other as the *straight-dough method.*

The long-process sponge method is employed when sufficient time can be allowed to permit the natural growth of the yeast. To make bread according to this process, start it in the evening by warming the liquid and dissolving the yeast and then adding these ingredients to the sugar, salt, and fat, which should first be placed in the mixing bowl. Stir this mixture well, and then add one-half of the quantity of flour that is to be used, stirring this also. Place this mixture, or sponge, as such a mixture is called, where it will remain warm, or at a temperature of from 65 to 70 degrees Fahrenheit, through the night. In the morning, stir the remaining flour into the sponge and knead for a few minutes the dough thus formed. When this is accomplished, put the dough in a warm place and allow it to rise until it doubles in bulk. When the dough is in this condition, it is ready to be kneaded again, after which it may be shaped into loaves, placed in the pans, allowed to double in bulk again, and finally baked.

The long-process straight-dough method is a shortened form

of the method just explained. It does away with the necessity of one kneading and one rising and consequently saves considerable time and labor. To make bread by this method, combine the ingredients in the evening as for the sponge method, but instead of adding only half of the flour, put all of it into the mixture, make a stiff dough at once, and knead. Then allow this to rise during the night, so that in the morning it can be kneaded again and put directly into the bread pans. After it rises in the pans until it doubles in bulk, it is ready to be baked.

The only disadvantage of the straight-dough method is that a stiff dough rises more slowly than a sponge, but since the entire night is given to the rising no difficulty will be experienced in carrying out this process. A point to remember, however, is that dough made according to this method must be kept warmer than that made by the sponge method.

QUICK PROCESS.—In the quick process of combining bread ingredients, there are also two methods of procedure—the *sponge method* and the *straight-dough method*. The chief differences between the methods of this process and those of the long process are in the quantity of yeast used and the length of time required for the bread to rise. More yeast must be used and much less time is required for the completion of the entire process. This shorter period of time is doubtless due to the fact that throughout the process, whether the straight-dough or the sponge method is followed, the mixture must be kept at a uniform temperature of about 90 degrees Fahrenheit.

The quick-process sponge method requires only about 5 hours for its completion, and the bread may be started at any time of the day that will allow this amount of time for carrying on the work. For this method, warm the ingredients and then combine the sugar, salt, fat, liquid, and dissolved yeast. Into this mixture, stir enough of the flour to make a sponge and put it where it will keep uniformly warm until it has about doubled in quantity and is full of bubbles. Then add the remainder of the flour, knead the mixture, and return the dough thus formed to a warm place. When the dough has doubled in bulk, remove it from the bowl to the kneading board, knead it slightly, and then shape it into loaves. Place these into the pans, and after allowing them to rise sufficiently, bake them.

The quick-process straight-dough method differs from the quick-process sponge method in that the entire amount of flour is added when the ingredients are first mixed, with the result that a stiff dough instead of a sponge is formed. As has already been learned, this stiff dough rises more slowly than a sponge, but it requires one rising less. It must be kept at a uniform temperature as much of the time as possible, so that the rising will not be retarded. When it has doubled in bulk, remove it from the bowl and knead it. Then shape it into loaves, place these in the pans, allow them to rise sufficiently, and proceed with the baking.

## CARE OF THE RISING DOUGH

PURPOSE OF RISING.—Rising is an important part of the process of bread making, no matter which method is employed. In a sponge, its purpose is to blend the ingredients after they have been mixed, as well as to permit the growth of the yeast; in a dough, after the gas has been evenly distributed by means of kneading, the purpose of rising is to permit the incorporation of a sufficient quantity of carbon dioxide to make the bread light when it is baked. As has just been explained, three risings are necessary in the sponge method of both the long and the short process, whereas only two are required in the straight-dough methods.

The last rising, or the one that takes place after the dough is shaped into loaves, is the one that affects the texture of the bread most, so that it should receive considerable attention. If the dough is not allowed to rise sufficiently at this time, the bread will be too fine in texture and will likely be heavy; and if it is permitted to rise too much, it will be coarse in texture. Allowance, however, should be made for the fact that the rising will continue after the bread has been placed in the oven.

TEMPERATURE FOR RISING.—As has been mentioned, the best results are obtained if the bread dough is kept at a uniform temperature throughout its rising. The temperature at which it rises most rapidly is about 86 degrees Fahrenheit; but, unless it can be watched closely, a better plan is to keep it, especially if the long process of bread making is followed, at a temperature that runs no higher than 80 degrees. Various methods of maintaining a uniform temperature have been devised, but the ones usually resorted to consist in placing the bowl containing the sponge or the dough in a bread raiser, a fireless cooker, or a vessel of hot water.

Bread raisers can be purchased, but if desired a simple bread-raising device may be constructed from a good-sized wooden box. To make such a device, line the box with tin or similar metal and fit it with a door or a cover that may be closed tight. Make a hole in one side of the box into which to insert a thermometer, and, at about the center of the box, place a shelf on which to set the bowl or pan containing the sponge or dough. For heating the interior, use may be made of a single gas burner, an oil lamp, or any other small heating device. This should be placed in the bottom of the box, under the shelf, and over it should be placed a pan of water to keep the air in the box moist, moist air being essential to good results. Where large quantities of bread must be baked regularly, such a device will prove very satisfactory. The temperature inside should be kept somewhere in the neighborhood of 95 to 105 degrees Fahrenheit if the bread is to rise rapidly; but it may be kept from 80 to 95 degrees if slower rising is desired.

Placing the bowl containing the dough mixture in a larger vessel of hot water is a simple and satisfactory way of obtaining a uniform temperature, being especially desirable for a sponge in the quick-process sponge method. The water in the large vessel should be at a temperature of about 110 to 115 degrees Fahrenheit. After the bowl of sponge or dough is placed in the water, the large vessel should be covered very carefully, so that the heat from the water will be retained. To maintain the temperature in the vessel and thus keep it right for the bread mixture, the hot water has to be replenished occasionally. If this is done, the sponge or dough will be maintained at a temperature of about 90 degrees and will therefore rise rapidly.

To insure the best results with the rising of bread mixtures, it is advisable, for the beginner at least, to use a thermometer for determining the temperature of air or water, as this instrument will save considerable time until experience in judging such matters has been gained. A Fahrenheit thermometer like that shown in Fig. 4 is the ideal kind for use in bread making. As an aid in this process, there are indicated in this illustration the temperature at which dough should be kept for rising and the temperature at which water should be kept outside the bowl to maintain a temperature of 75 to 90 degrees in the dough when the plan mentioned in Art. 48 for keeping dough at a uniform temperature is followed. In addition, the oven temperatures for baking bread and rolls, which are explained later, are also shown. The temperature of water can, however, be determined fairly accurately with the hands. If it feels very warm but does not burn the hand, it may be considered at about a temperature of 110 to 115 degrees.

In order to prevent the formation of a hard surface on the dough, the bowl in which it rises should be kept tightly covered. A further means of preventing this condition consists in oiling the surface of the dough; that is, brushing it lightly with melted fat. In case a crust does form, it should be well moistened with water or milk and allowed to soften completely before the next kneading is begun.

TIME REQUIRED FOR RISING.—No definite rule can be given for the length of time required for dough to rise, for this depends entirely on the activity of the yeast. If the yeast is active, the dough will rise quickly; but if it is not of good quality or if it has been killed or retarded in its growth by improper handling, the dough will rise slowly. Usually, dough should be allowed to rise until it has doubled in bulk. A good way in which to determine when this takes place is to put a small piece of the dough in a glass, such as a measuring glass, a tumbler, or a jelly glass, and mark on this glass where the dough should come when it has increased to twice its size. This glass set beside the vessel containing the dough will show when it has risen sufficiently.

## KNEADING THE DOUGH

PURPOSE OF KNEADING.—As has been pointed out, it is necessary to knead dough one or more times in the making of bread, the number of kneadings depending on the method that is employed. The purpose of kneading is to work the dough so as to distribute evenly the gas that is produced by the yeast, to increase the elasticity of the gluten, and to blend the ingredients. It is a very important part of the work of bread making, for to a great extent it is responsible for the texture of the finished product. At first, kneading may be found to be somewhat difficult, but the beginner need not become discouraged if she is not proficient at once, because the skill that is necessary to knead the bread successfully comes with practice. So that the best results may be attained, however, it is advisable that the purpose for which the kneading is done be kept constantly before the mind during the process.

KNEADING MOTIONS.—Several motions are involved in the kneading of bread, and these are illustrated in Figs. 7 to 10. In order to carry out the kneading process, first cover lightly with flour the surface on which the kneading is to be done; this may be a suitable table top or a molding board placed on a table. Then remove the dough from the mixing bowl with the aid of a case knife or a spatula, in the manner shown in Fig. 6, and place it on the floured surface. Sift a little flour over the dough, so that it appears as in Fig. 7, and flatten it slightly by patting it gently. Next, with the fingers placed as shown in Fig. 8, take hold of the edge of the mass at the side farthest from you and fold the dough over the edge nearest you, as Fig. 9 illustrates. Then work the dough with a downward pressure and, as indicated in Fig. 10, push it out with the palms of the hands. With the motion completed, turn the entire mass around and knead it in the same way in another direction. Continue the kneading by repeating these motions until the dough has a smooth appearance, is elastic, does not stick to either the hands or the board, and rises quickly when it is pressed down.

To prevent the dough from sticking to the hands and the board, flour should be added gradually during the process of kneading, but care should be taken not to use too much flour for this purpose. The lightness and sponginess of the finished loaf depend largely on the quantity of flour used at this time, so that if the dough is made too stiff with flour, the bread will be hard and close after it is baked. As soon as the dough can be kneaded without its sticking to either the hands or the board, no more flour need be added; but, in case too much flour is used, the dough may be softened by means of milk or water. Such dough, however, is not so satisfactory as that which does not have to be softened.

## SHAPING THE DOUGH INTO LOAVES

After the dough is properly kneaded in the manner just explained, it is placed in the mixing bowl and allowed to rise again. When it has risen sufficiently for the last time, depending on the process employed, it should be kneaded again, if it must be reduced in size, and then shaped into loaves and put in the pans. Here, again, much care should be exercised, for the way in which bread is prepared for the pans has much to do with the shape of the loaf after it is baked.

In order to shape the dough into loaves, first loosen it from the sides of the mixing bowl, using a knife or a spatula for this purpose, and then turn it out on a flat surface on which flour has been sprinkled, as in preparing for kneading. Knead the dough a little, and then cut it into pieces that will be the correct size for the pans in which the loaves are to be baked. Dust each piece with a small quantity of flour and knead it until the large bubbles of gas it contains are worked out and it is smooth and round. In working it, stretch the under side, which is to be the top of the loaf, and form it into a roll that is as long and half as high as the pan and as thick at each end as in the center.

## BAKING THE BREAD

PURPOSE OF BAKING.—The various processes in the making of bread that have been considered up to this point may be successfully carried out, but unless the baking, which is the last step, is properly done, the bread is likely to be unpalatable and indigestible. Much attention should therefore be given to this part of the work. So that the best results may be obtained, it should be borne in mind that bread is baked for the purpose of killing the ferment, rupturing the starch grains of the flour so that they become digestible, fixing the air cells, and forming a nicely flavored crust. During the process of baking, certain changes take place in the loaf. The gluten that the dough contains is hardened by the heat and remains in the shape of bubbles, which give the bread a porous appearance; also, the starch contained in the dough is cooked within the loaf, but the outside is first cooked and then toasted.

OVEN TEMPERATURE FOR BAKING.—In baking bread, it is necessary first to provide the oven with heat of the right temperature and of sufficient strength to last throughout the baking. The usual oven temperature for successful bread baking is from 380 to 425 degrees Fahrenheit, but in both the first and the last part of the baking the heat should be less than during the middle of it. An oven thermometer or an oven gauge is a very good means of determining the temperature of the oven. But if neither of these is available the heat may be tested by placing in the oven a white cracker, a piece of white paper, or a layer of flour spread on a shallow tin pan. If any one of these becomes a light brown in 5 minutes, the oven is right to commence baking. Every precaution should be taken to have the oven just right at first, for if the bread is placed in an oven that is too hot the yeast plant will be killed immediately and the rising consequently checked. Of course, the bread will rise to some extent even if the yeast plant is killed at once, for the carbon dioxide that the dough contains will expand as it becomes heated and will force the loaf up; but bread baked in this way is generally very unsatisfactory, because a hard crust forms on the top and it must either burst or retard the rising of the loaf.

If the heat is not sufficient, the dough will continue to rise until the air cells run together and cause large holes to form in the loaf. In an oven that is just moderately hot, or has a temperature of about 400 degrees, the yeast plant will not be killed so quickly, the dough will continue to rise for some time, and the crust of the bread should begin to brown in about 15 minutes.

TIME FOR BAKING AND CARE OF BREAD IN OVEN.—The time required for baking bread and the care it should receive in the oven are also important matters to know. How long the bread should bake depends on the size of the loaf. Under proper oven temperature, a small loaf, or one made with 1 cupful of liquid, ought to bake in from 50 minutes to 1 hour, while a large loaf requires from 1-1/2 to 2 hours. As has been explained, the loaf should begin to brown, or have its crust formed, in about 15 minutes after it is placed in the oven, and the baking should proceed rather slowly.

To get the best results in baking, the pans should be placed so that the air in the oven will circulate freely around them. If they are so placed that the loaves touch each other or the sides of the oven, the loaves will rise unevenly and consequently will be unsightly in shape, like those shown in Figs. 14 and 15. If the loaves rise higher on one side than on the other, even when the pans are properly placed, it is evident that the heat is greater in that place than in the other parts of the oven and the loaves should therefore be changed to another position. Proper care given to bread while baking will produce loaves that are an even brown on the bottom, sides, and top and that shrink from the sides of the pan.

## CARE OF BREAD AFTER BAKING

As soon as the bread has baked sufficiently, take it from the oven, remove the loaves from the pans, and place them to cool where the air may circulate freely around them. A bread rack, or cake cooler, like the one on which the loaf rests, is very satisfactory for this purpose, but if such a device is not available, the loaves may be placed across the edges of the empty pans so that nearly the entire surface is exposed.

Whichever plan is adopted, it should be remembered that the bread must be carefully protected from dust and flies. Bread should never be permitted to remain in the pans after it has been baked nor to cool on a flat surface; neither should the loaves be wrapped while they are warm, because the moisture will collect on the surface and the bread will not keep so well.

After the loaves have become sufficiently cool, place them in the receptacle in which they are to be kept.

This should have been previously washed and dried and then allowed to stand in the sunshine, so as to be free from mold or any substance that will taint or otherwise injure the bread. After the loaves have been put into it, keep it well covered and allow no stale crumbs nor pieces of bread to collect. To keep such a receptacle in good condition, it should be scalded and dried every 2 or 3 days.

## SCORING BREAD

OBJECT OF SCORING BREAD.—By the *scoring* of bread is meant simply the judging of its qualities. Persons who understand what good bread is agree very closely on the qualities that should characterize it, and they make these qualities a standard by which any kind of bread may be scored, or judged.

Those who are not proficient in the making of bread, as well as those who have had very little experience, will do well to have their bread judged by experts or to learn how to score it themselves. By following this plan, they will be able to find out the good and bad points of their bread and then, by ascertaining the causes of any poor qualities, will be in a position to make improvements. So that the beginner may learn how to judge the qualities of her bread, she should study carefully the accompanying score card and its explanation.

| SCORE CARD | |
|---|---|
| External Appearance: | |
| Shape | 5 % |
| Size | 2 % |
| Crust: | |
| Shade | 2 % |
| Uniformity of Color | 2 % |
| Character | 2 % |
| Depth | 2-8 % |
| Lightness | 20 % |
| Internal Appearance: | |
| Even distribution of gas | 10 % |
| Moisture | 5 % |
| Elasticity | 5 % |
| Color | 15 % |
| Flavor | 30 % |
| Total | 100 % |

EXPLANATION OF SCORE CARD.—A study of the score card will reveal that a certain number of points are given to a loaf of bread for appearance, both external and internal, for lightness, and for flavor. To determine these qualities best, allow the loaf to cool thoroughly after baking. Then consider the various points, and decide how nearly perfect the loaf is in respect to each one of them. Add the numbers that are determined upon, and the result obtained will show how the bread scores.

The *shape* of the loaf, in order to be perfect and to score 5, should be uniform and symmetrical.

The *size* of the loaf, for which a score of 2 is given, is determined from the standpoint of thorough baking. The exact size that a loaf must be is a rather difficult thing to state, because the sizes vary considerably, but a loaf of an ungainly size should be guarded against, for it would not score well. Bread made in pans of the size already mentioned would score high with regard to size.

The *crust*, whose combined characteristics score 8, should be a golden brown in color in order to receive the score of 2 for its *shade*. A pale loaf or one baked too brown would not receive full credit. If the required color extends uniformly over the entire loaf, the bottom and the sides, as well as the top, 2 more is added to the score of the crust for *uniformity of color*. After these points are scored, a slice of bread should be cut from the loaf in order that the remaining points may be scored. As fresh bread does not cut easily, and as a well-cut slice must be had for this purpose, special care must be taken to obtain the slice. Therefore, sharpen a large knife and heat the blade slightly by holding it near a flame; then cut a slice at least 1/2 inch thick from the loaf before the blade has had time to cool. With such a slice cut, the *character* of the crust, by which is meant its toughness or its tenderness, may be determined. A score of 2 is given if it is of sufficient tenderness or is devoid of toughness. The *depth* of the crust, which depends on the amount of baking the loaf has had, receives a score of 2 if it is perfect. A deep crust, which is the preferred kind, is produced by long, slow baking; bread that is baked only a short time has a thin crust, which is not so desirable and would not score so high.

The *lightness* of the bread can easily be scored when the bread is cut. It is judged by the size of the holes, and if it is perfect it receives a score of 20. If the bread is not light enough, the holes will be small and the bread will feel solid and unelastic; if it is too light, the holes will be large and coarse.

The internal appearance, which is scored next, includes several characteristics. For the *even distribution of gas*, which is determined by the uniformity of the holes, 10 points are given. If the kneading has been done right and the bread has risen properly, the gas will be distributed evenly through the loaf, with the result that the holes, which make the bread porous, will be practically the same throughout the entire loaf. Such a texture is better than that of a loaf that has some large and some small holes. The *moisture* in the bread, which receives 5 if it is of the right amount, is tested by pinching a crumb between the fingers. If the crumb feels harsh and dry, the bread is not moist enough, and if it feels doughy, the bread is too moist. The *elasticity*, for which 5 is given, is determined by pressing the finger gently into a cut place in the loaf. The bread may be considered to be elastic if it springs back after the finger is removed and does not break nor crumble. As compared with cake, bread is always more elastic, a characteristic that is due to the quantity of gluten it contains. Still it should be remembered that the elasticity must not amount to toughness, for if it does the quality of the bread is impaired. To score 15 for *color*, the inside of the loaf should be of an even, creamy white. A dull white or gray color would indicate that flour of a poor quality had been used, and dark or white streaks in the bread would denote uneven mixing and insufficient kneading.

The last thing to be scored, namely, the *flavor*, merits 30 points. To determine this characteristic, chew a small piece of bread well. If it is not sour nor musty, has a sweet, nutty flavor, and shows that the correct amount of salt and sugar were added in the mixing, it may receive a perfect score.

## USE OF THE BREAD MIXER

The advantage of a bread mixer in bread making is that it practically does away with hand mixing and kneading; however, all the other steps described are the same, depending on the process used. As has been mentioned, the housewife who bakes such a small quantity as three or four loaves of bread can get along very well without a bread mixer; at least, for so few loaves a bread mixer does not seem so necessary as when six or more loaves are to be made at one time, when it is a decided convenience. However, bread mixers can be had in various sizes to meet the requirements of the housewife.

In using a bread mixer like that described in *Essentials of Cookery*, Part 2, the ingredients are placed in the mixer and thoroughly mixed together by turning the handle, and after the sponge or the dough has risen, the kneading is performed by again turning the handle. The amount of turning to be done is, of course, regulated by the ingredients and the method that is followed.

In addition to the bread mixer mentioned, there is another convenient type that is constructed in two parts, the top part having a sifter in its bottom, through which the flour or other dry ingredients are sifted. The sifting is done with a crank, which also operates a shaft to which is attached a number of knives extending in different directions. These knives accomplish the mixing and the kneading. The bread is allowed to rise in the lower part of the bread mixer, the top part being removed after the mixing and sifting have been accomplished.

Any of the bread-making methods described may be used with the bread mixer without change in the process, and no kneading need be done by hand except a sufficient amount to shape the loaves after the last rising and before they are placed in the pans.

## SERVING BREAD

Bread is one of the foods that every one takes so much as a matter of course that little thought is given to its serving. Of course, it does not offer so much opportunity for variety in serving as do some foods; yet, like all other foods, it appeals more to the appetites of those who are to eat it if it is served in an attractive manner. A few ideas as to the ways in which it may be served will therefore not be amiss.

As fresh bread is not easily digested, it should not usually be served until it is at least 24 hours old. Before it is placed on the table, it should be cut in slices, the thickness of which will depend on the preference of the persons who are to eat it. If the loaf is large in size, the pieces should be cut in two, lengthwise of the slice, but in the case of a small loaf the slices need not be cut.

Various receptacles for placing bread and rolls on the table, such as a bread boat, a bread plate, and a bread basket, are also used to add variety in serving. Whichever of these is selected, it may be improved in appearance by the addition of a white linen doily. For rolls, a hot-roll cover is both convenient and attractive. Sometimes, especially when a large number of persons are to be served, a roll is placed between the folds of each person's napkin before they are seated at the table.

Occasionally bread becomes stale before it is needed on the table. Such bread, however, should not be discarded, especially if the loaves are uncut. Uncut loaves of this kind may be freshened by dipping them quickly into boiling water and then placing them in a very hot oven until their surface becomes dry. If desired, slices of bread that have become stale may be steamed in order to freshen them; but unless great care is taken in steaming them the bread is liable to become too moist and soggy.

## BREAD RECIPES

In order that the beginner may bring into use the bread-making principles and directions that have been set forth, and at the same time become familiar with the quantities of ingredients that must be used, there are here given a number of recipes for the making of bread. These recipes include not only white bread-that is, bread made from white flour—but whole-wheat, graham, rye, and corn bread, as well as bread in which fruit and nuts are incorporated. Before these recipes are taken up, though, it will not be amiss to look further into the various ingredients used in the making of bread.

The fat used in bread making may vary in both quantity and kind. For instance, if less than 2 tablespoonfuls is called for in a recipe, this amount may be decreased; but it is not well to increase the amount to any extent. Likewise, the fat may be of any kind that will not impart a disagreeable flavour to the finished product. It may be left-over chicken fat, clarified beef fat, lard, butter, cooking oil, or any mixture of clear, fresh fats that may be in supply.

The sweetening for bread is, as a rule, granulated sugar, although sirup, molasses, brown sugar, or white sugar of any kind may be employed. Sweetening is used merely to give a slightly sweet flavour to the bread, and the kind that is used is of slight importance.

The liquid, as has been stated, may be water or milk or any proportion of both. The milk that is used may be either whole or skim. In addition to these two liquids, the whey from cottage cheese or the water in which rice, macaroni, or potatoes have been cooked should not be overlooked. Potato water in which a small quantity of potato may be mashed serves as a yeast aid, as has been pointed out. Therefore, whenever, in a bread recipe, liquid is called for and the kind to be used is not stated specifically, use may be made of any of the liquids that have been mentioned.

The quantity of flour required for a bread recipe will depend entirely on the kind of flour that is to be used, bread flour having a much greater absorbing power for liquid than has pastry or blend flour. When, in the process of mixing the bread, the sponge is stiffened by adding the remaining flour to it, the last cupful or two should be added cautiously, in order not to make the mixture too stiff. In some instances, more flour than the recipe calls for may be required to make the dough of the right consistency. The amount can be determined only by a knowledge of what this consistency should be, and this will be easily acquired with practice in bread making. The beginner will find it a good plan to begin making bread entirely of white flour, for the reason that it is easier to determine the consistency of the dough mixture at various stages, as well as during the kneading, if there is no coarse material, such as bran, corn meal, nuts, fruits, etc., in the dough. Later, when a definite knowledge along this line has been acquired, one after the other of the bread recipes should be tried. They are no more difficult to carry out than the recipes for white bread; indeed, the woman who has had experience in bread making will find that she will be equally successful with all of them.

WHITE BREAD.—Bread made from white flour, which is commonly referred to as *white bread*, is used to a much greater extent than any other kind, for it is the variety that most persons prefer and of which they do not tire quickly. However, white bread should not be used to the exclusion of other breads, because they are of considerable importance economically. This kind of bread may be made by both the quick and the long processes, for the ingredients are the same, with the exception of the quantity of yeast used. The amounts given in the following recipes are sufficient to make two large loaves or three small ones, but, of course, if more bread is desired, the quantity of each ingredient may be increased proportionately.

### Whole-wheat Bread—Quick Process

(*Sufficient for Two Small Loaves*)

3 Tb. fat

1/4 c. brown sugar

1 Tb. salt

1 cake compressed yeast

3 c. lukewarm liquid

8 c. whole-wheat flour

1 c. white flour for kneading

Place the fat, the sugar, and the salt in the mixing bowl and add the yeast cake dissolved in a little of the liquid. Add the remainder of the liquid, and then stir in half or all of the flour, according to whether the sponge or the straight-dough method is preferred. Then proceed according to the directions previously given for making bread by the quick process.

The long process may also be followed in making whole-wheat bread, and if it is, only one-half the quantity of yeast should be used.

GRAHAM BREAD.—To lend variety to the family diet, frequent use should be made of graham bread, which contains even more bulk and mineral salts than whole-wheat bread. In bread of this kind, both graham and white flour are used. Since graham flour is very heavy, it prevents the bread from rising quickly, so the bread is started with white flour. The accompanying recipe contains quantities for the short process, although it may be adapted to the long process by merely using one-half the amount of yeast.

**Graham Bread**

(*Sufficient for Two Loaves*)

2 Tb. fat

1/4 c. brown sugar

2 tsp. salt

1 cake compressed yeast

2 c. lukewarm liquid

2 c. white flour

3 c. graham flour 1 c. white flour additional for kneading

Put the fat, the sugar, and the salt in the mixing bowl, and to them add the yeast that has been dissolved in a little of the liquid. Pour over these ingredients the remainder of the liquid and stir in the white flour. When the mixture is to be made stiff, add the graham flour.

Then knead the dough, let it rise, knead again, place it in greased pans, let rise, and bake. A point to be remembered in the making of graham bread is that sifting removes the bran from graham flour, and if lightness is desired, the flour may be sifted and the bran then replaced.

GRAHAM BREAD WITH NUTS.—To increase the food value of graham bread, nuts are sometimes added. This kind of bread also provides an agreeable variety to the diet. The following recipe is intended to be carried out by the short process, so that if the long process is desired the quantity of yeast must be reduced.

### White Bread—Long Process

(*Sufficient for Two Large or Three Small Loaves*)

2 Tb. fat

2 Tb. sugar

1/2 cake compressed yeast, or 1 cake dried yeast 1 Tb. salt 1 qt. lukewarm liquid

3 qt. flour

1 c. flour additional for kneading

Put into the mixing bowl the fat, the sugar, the salt, and the yeast that has been dissolved in a little of the lukewarm liquid. Add the remainder of the liquid and stir in half of the flour. Place this sponge where it will rise overnight and will not become chilled. In the morning, add the remainder of the flour, stirring it well into the risen sponge, and knead the dough thus formed. Allow it to rise until it has doubled in bulk and then knead it again. After it is properly kneaded, shape it into loaves, place them in greased pans, let them rise until they have doubled in bulk, and then bake them.

Combining the ingredients in the manner just mentioned is following the sponge method of the long process. By adding all instead of half of the flour at night, the straight-dough method of this process may be followed.

### Graham Bread With Nuts

(*Sufficient for Two Loaves*)

1 cake compressed yeast

2 c. lukewarm liquid 1/4 c. molasses 2 Tb. fat

1 Tb. salt

2 c. white flour

4 c. graham flour

1-1/2 c. chopped nuts

1 c. white flour additional for kneading

Dissolve the yeast in a little of the lukewarm liquid and mix it with the molasses, fat, and salt. Add the remaining liquid and the white flour. Let this sponge rise until it is light. Then stir in the graham flour, adding the nuts while kneading. Let the dough rise until it doubles in bulk. Shape into loaves, place it in the greased pans, and let it rise until it doubles in size. Bake for an hour or more, according to the size of the loaves.

WHOLE-WHEAT FRUIT BREAD.—A very delicious whole-wheat bread is produced by combining fruit, which, besides improving the flavour, adds to the food value of the bread. Thin slices of this kind of bread spread with butter make excellent summer sandwiches. If the short process is employed, the amounts specified in the following recipe should be used, but for the long process the quantity of yeast should be decreased.

**Whole-wheat Fruit Bread**

(*Sufficient for Three Small Loaves*)

1 yeast cake

2 c. lukewarm liquid 2 Tb. fat

1/4 c. brown sugar stoned, chopped dates 2 tsp. salt

6 c. whole-wheat flour 1-1/2 c. seeded raisins or stoned, chopped dates

1 c. white flour for kneading

Dissolve the yeast cake in a little of the lukewarm liquid and add it to the fat, sugar, and salt that have been put into the mixing bowl. Pour in the remainder of the liquid and add half or all of the flour, depending on the bread-making method that is followed. Stir in the fruit before all the flour is added and just before the dough is shaped into loaves. After it has risen sufficiently in the greased pans, proceed with the baking.

BRAN BREAD.—Bread in which bran is used is proportionately a trifle lower in food value than that in which whole wheat or white flour is used. However, it has the advantage of an additional amount of bulk in the form of bran, and because of this it is a wholesome food.

**Bran Bread**

(*Sufficient for Two Loaves*)

2 c. milk

6 Tb. molasses

1-1/2 tsp. salt

1/2 yeast cake

1/4 c. lukewarm water

2 c. white flour

4 c. graham flour

1 c. sterilized bran

1 c. white flour additional for kneading

Scald the milk and to it add the molasses and salt. When this is lukewarm, add to it the yeast cake dissolved in the lukewarm water, as well as the white flour and 1 cupful of the graham flour. Cover this mixture and let it rise.

When it has risen sufficiently, add the bran and the rest of the graham flour and knead. Cover this dough, and let it rise until it doubles in bulk. Then shape it into loaves, place it in the greased pans, let it rise again until it doubles in bulk, and bake in a hot oven.

RYE BREAD.—Rye bread has a typical flavour that many persons enjoy. When rye flour is used alone, it makes a moist, sticky bread; therefore, in order to produce bread of a good texture, wheat flour must be used with the rye flour. The recipe here given is for the short process of bread making, but by reducing the quantity of yeast it may be used for the long process.

**Rye Bread**

(*Sufficient for Three Loaves*)

2 Tb. fat

1 Tb. salt

2 Tb. sugar 1 cake compressed yeast

3 c. lukewarm liquid 6 c. rye flour

4 c. white flour 1 c. white flour additional for kneading

Into the mixing bowl, put the fat, the salt, the sugar, and the yeast that has been dissolved in a small quantity of the lukewarm liquid. Then stir in the flour, one-half or all of it, according to whether the sponge or the straight-dough method is followed. When the dough is formed, allow it to rise until it doubles in bulk; then knead it and shape it into loaves for the greased pans.

When these have risen until they are double in size and therefore ready for the oven, glaze the surface of each by brushing it with the white of egg and water and put them in the oven to bake. If desired, caraway seed may be added to the dough when it is formed into loaves or simply sprinkled on the top of each loaf. To many persons the caraway seed imparts a flavour to the bread that is very satisfactory.

CORN BREAD.—Corn meal is sometimes combined with wheat flour to make corn bread. Such a combination decreases the cost of bread at times when corn meal is cheap. Bread of this kind is high in food value, because corn meal contains a large proportion of fat, which is more or less lacking in white flour. The following recipe is given for the short process, but it may be used for the long process by merely decreasing the quantity of yeast.

**Corn Bread**

(*Sufficient for Two Loaves*)

1 yeast cake

2 c. lukewarm liquid 2 tsp. salt

1 Tb. sugar

2 Tb. fat

4-1/2 c. white flour 2 c. corn meal 1 c. white flour additional for kneading

Put the yeast to soak in 1/4 cupful of warm water and let it dissolve. Heat the liquid and cool it to lukewarm, and then add to it the salt, the sugar, the dissolved yeast, and the melted fat. Make a sponge with some of the flour and let it rise until it doubles in bulk. Then make a dough with the corn meal and the remaining flour. Knead the dough, let it rise again, and form it into loaves. Let these rise in the greased pans until they double in bulk; then bake about 45 minutes.

RICE BREAD.—Very often variety is given to bread by the addition of rice, which imparts an unusual flavour to bread and effects a saving of wheat flour. Oatmeal and other cereals may be used in the same way as rice, and bread containing any of these moist cereals will remain moist longer than bread in which they are not used.

**Rice Bread**

(*Sufficient for Three Loaves*)

1/2 c. uncooked rice

1-1/2 c. water

1 Tb. salt

1 Tb. sugar

1 Tb. fat

1/2 yeast cake

1 c. lukewarm liquid

6 c. white flour

1 c. white flour additional for kneading

Steam the rice in a double boiler in 1 and a half cupfuls of water until it is soft and dry. Add the salt, sugar, and fat, and allow all to become lukewarm. Dissolve the yeast in the lukewarm liquid, and add it to the rice. Put all in the mixing bowl, stir in 2 cupfuls of flour, and allow the mixture to become very light. Add the remainder of the flour and knead lightly. Let the dough rise until it has doubled in bulk and knead to reduce the quantity. Place in greased pans. When the loaves have risen sufficiently, bake for about 50 minutes.

SALT-RISING BREAD.—Recipes for bread would be incomplete if mention were not made of salt-rising bread. Such bread differs from ordinary bread in that the gas that causes the rising is due to the action of bacteria. Salt-rising bread is not universally popular, yet many persons are fond of it. Its taste is very agreeable, and, as a rule, its texture is excellent; however, it always has an unpleasant odour. The method given in the accompanying recipe for salt-rising bread differs in no way from the usual method of making it. It is very necessary that the first mixture of corn meal, salt, sugar, and milk be kept at a uniformly warm temperature in order to induce bacteria to grow. Any failure to make such bread successfully will probably be due to the violation of this precaution rather than to any other cause.

**Salt-rising Bread**

(*Sufficient for Two Loaves*)

1 c. fresh milk 1/4 c. corn meal

1 tsp. salt

2 tsp. sugar 2 c. lukewarm water

7 c. white flour

1/2 c. white flour additional for kneading

Scald the milk and pour it over the corn meal, salt, and sugar. Allow this mixture to stand in a warm place for several hours or overnight, when it should be light. To this batter add the warm water and enough flour to make a drop batter. Allow this to stand in a warm place until it is light; and then add the remainder of the flour so as to make a dough, and knead. Allow this to rise, shape it into loaves, put it in pans, let it rise again, and bake.

## RECIPES FOR ROLLS, BUNS, AND BISCUITS

While the preceding recipes call for bread in the form of loaves, it should be understood that bread may be made up in other forms, such as rolls, buns, and biscuits. These forms of bread may be made from any of the bread recipes by adding to the mixture shortening, sugar, eggs, fruit, nuts, spices, flavoring, or anything else desirable. Since these things in any quantity retard the rising of the sponge or dough, they should be added after it has risen at least once.

Rolls, buns, and biscuits may be made in various shapes. To shape them, the dough may be rolled thin and then cut with cutters, or the pieces used for them may be pinched or cut from the dough and shaped with the hands. After they are shaped, they should be allowed to rise until they double in bulk. To give them a glazed appearance, the surface of each may be brushed before baking with milk, with white of egg and water, or with sugar and water. Butter is also desirable for this purpose, as it produces a crust that is more tender and less likely to be tough. Rolls, buns, or biscuits may be baked in an oven that has a higher temperature than that required for bread in the form of loaves, and only 15 to 20 minutes is needed for baking them. If such forms of bread are desired with a crust covering the entire surface, they must be placed far enough apart so that the edges will not touch when they are baking.

So that experience may be had in the preparation of rolls, buns, and biscuits there are given here several recipes that can be worked out to advantage, especially after proficiency in bread making has been attained.

PARKER HOUSE ROLLS.—Of the various kinds of rolls, perhaps none meets with greater favor than the so-called Parker House rolls, one of which is shown at *a*, Fig. 19. Such rolls may be used in almost any kind of meal, and since they are brushed with butter before they are baked, they may be served without butter, if desired, in a meal that includes gravy or fat meat.

### Parker House Rolls

(*Sufficient for 3 Dozen Rolls*)

1 cake compressed yeast

1 pt. lukewarm milk 4 Tb. fat

2 Tb. sugar 1 tsp. salt

3 pt. white flour 1 c. white flour additional for kneading

Dissolve the yeast in some of the lukewarm milk. Pour the remainder of the warm milk over the fat, sugar, salt, and dissolved yeast, all of which should first be put in a mixing bowl. Stir into these ingredients half of the flour, and beat until smooth. Cover this sponge and let it rise until it is light. Add the remainder of the flour, and knead until the dough is smooth and does not stick to the board. Place the dough in a greased bowl, and let it rise again until it doubles in bulk. Roll the dough on a molding board until it is about 1/4 inch thick. Then cut the rolled dough with a round cutter; brush each piece with soft butter; mark it through the center, as at *b*, Fig. 19, with the dull edge of a kitchen knife; and fold it over, as at *c*. Place the pieces of dough thus prepared in shallow pans, about 1 inch apart, and let them rise until they are light, when each roll will appear like that shown at *d*. Then bake them in a hot oven for about 15 minutes.

DINNER ROLLS.—As their name implies, dinner rolls are an especially desirable kind of roll to serve with a dinner. They should be made small enough to be dainty, and as an even, brown crust all over the rolls is desirable they should be placed far enough apart in the pans to prevent them from touching one another. If they are placed as in (*b*), that is, close together, only part of the crust will be brown. When made according to the accompanying recipe, dinner rolls are very palatable.

### Dinner Rolls

(*Sufficient for 1-1/2 Dozen Rolls*)

1 cake compressed yeast

1 c. lukewarm milk

2 Tb. sugar

2 Tb. fat 1 tsp. salt

3 c. white flour 1 egg white 1/2 c. white flour additional for kneading

Dissolve the yeast in some of the lukewarm milk. Put the sugar, fat, salt, and dissolved yeast in the mixing bowl, and pour the remainder of the milk over these ingredients. Stir half of the flour into this mixture and allow the sponge to rise. When it is light, add the egg white, which should first be beaten, and the remainder of the flour, and then knead the dough. Let the dough rise until it doubles in bulk. Roll out the dough until it is 1/2 inch thick, and then cut out the rolls with a small round cutter. Place these in a shallow pan and let them rise until they are light. Then glaze each one with the white of egg to which is added a little water and bake them in a hot oven for about 15 minutes.

LUNCHEON ROLLS.—If rolls smaller than dinner rolls are desired, luncheon rolls will undoubtedly be just what is wanted. Since these are very small, they become thoroughly baked and are therefore likely to be even more digestible than bread or biscuit dough baked in a loaf. For rolls of this kind, the following recipe will prove satisfactory:

## Luncheon Rolls

(*Sufficient for 2 Dozen Rolls*)

1 cake compressed yeast 1-1/4 c. lukewarm milk

2 Tb. sugar 2 Tb. fat 1 tsp. salt

4 c. white flour

1 egg white

1/2 c. white flour additional for kneading

Combine the ingredients in the manner directed for making dinner rolls. Shape the dough into biscuits the size of a small walnut, place them in a shallow pan, spacing them a short distance apart, and let them rise until they are light. Next, brush the tops of them with melted butter, and then bake them in a hot oven for about 15 minutes.

WHOLE-WHEAT ROLLS.—Rolls made of whole-wheat flour are not so common as those made of white flour, and for this reason they appeal to the appetite more than ordinary rolls. Whole-wheat rolls have the same advantage as bread made of whole-wheat flour, and if they are well baked they have a crust that adds to their palatableness.

### Whole-wheat Rolls

(*Sufficient for 3 Dozen Rolls*)

1 pt. lukewarm milk 1 cake compressed yeast

1 tsp. salt

3 Tb. sugar

4 Tb. fat

2 c. white flour

4 c. whole-wheat flour

1/2 c. white flour additional for kneading

Set a sponge with the lukewarm milk, in which are put the yeast cake, salt, sugar, fat, and white flour. Allow this to become very light, and then add the whole-wheat flour. Knead this dough and allow it to double in bulk. Then shape it into rolls, allow them to rise, and bake for 15 to 20 minutes.

GRAHAM NUT BUNS.—Buns made of graham flour and containing nuts are not only especially delightful in flavour, but highly nutritious. Because they are high in food value, they may be served with a light meal, such as lunch or supper, to add nutrition to it. The recipe here given will result in excellent buns if it is followed closely.

### Graham Nut Buns

(*Sufficient for 3 Dozen Buns*)

1 cake compressed yeast

2 c. lukewarm milk 4 Tb. brown sugar 2 tsp. salt 2 Tb. fat

2-1/2 c. white flour 1 egg

1 c. chopped nuts 3-1/2 c. graham flour 1 c. white flour additional for kneading

Dissolve the yeast in a little of the lukewarm milk. Place the sugar, salt, fat, and dissolved yeast in the mixing bowl and add the remainder of the warm milk. Stir in the white flour and let the sponge thus formed rise. Then add the egg, which should first be beaten, the nuts, and the graham flour. Knead the dough and shape it into buns. Let these rise and then bake them in a hot oven for about 15 minutes.

NUT OR FRUIT BUNS.—Nuts or fruit added to buns made of white flour provide more mineral salts and bulk, substances in which white flour is lacking. Buns containing either of these ingredients, therefore, are especially valuable in the diet. Besides increasing the food value of the buns, nuts and fruit improve the flavour and make a very palatable form of bun. Buns of this kind are made as follows:

**Nut Or Fruit Buns**

(*Sufficient for 2 Dozen Buns*)

4 Tb. sugar 1 Tb. fat 1 tsp. salt

1 cake compressed yeast

1 c. lukewarm milk

3 c. white flour

3/4 c. chopped nuts or raisins

1 c. white flour additional for kneading

Add the sugar, fat, and salt to the yeast dissolved in a little of the milk. Then stir in the remainder of the milk and half of the flour. Allow this sponge to rise until it is very light, and then add the remainder of the flour and the nuts or the raisins. Knead at once and form into buns. Let these rise until they are light. Then moisten them with milk and sprinkle sugar over them before placing them in the oven. Bake for about 15 minutes.

SWEET BUNS.—Persons who prefer a sweet bun will find buns like those shown in Fig. 21 and made according to the following recipe very much to their taste. The sweetening, eggs, and lemon extract used in this recipe give to the white buns a delightful flavour and help to lend variety to the usual kind of bun.'

**Sweet Buns**

(*Sufficient for 1-1/3 Dozen Buns*)

1 cake compressed yeast

1 c. lukewarm scalded milk 1/4 c. sugar

2 Tb. fat 1 tsp.

1 tsp. salt 3-1/2 c. white flour

2 eggs

1 tsp. lemon extract 1 c. white flour additional for kneading

Dissolve the yeast in a small amount of the lukewarm milk and add it to the sugar, fat, salt, and remaining milk in the mixing bowl. Stir into this mixture half of the flour, beat well, and let the sponge rise until it is light. Add the eggs, which should first be beaten, the lemon extract, and the remaining flour. Knead until the dough is smooth. Let the dough rise again and then shape it into rolls. Allow these to rise, and then bake them in a hot oven for about 15 minutes.

COFFEE CAKE.—When an especially good kind of biscuit that can be served for breakfast and eaten with coffee is desired, coffee cake made according to the following recipe should be used. Cinnamon sprinkled over the top of such cake imparts a very pleasing flavour, but if more of this flavour is preferred 1 teaspoonful of cinnamon may be mixed with the dough.

## Coffee Cake

(*Sufficient for One Cake*)

1 cake compressed yeast 1/2 c. lukewarm milk

1 Tb. sugar 1/2 tsp. salt

2 c. white flour

1 egg

2 Tb. fat 1/4 c. brown sugar

1 1/2 c. white flour additional for kneading

Dissolve the yeast in the lukewarm milk and add the sugar and the salt. Stir in 1 cupful of flour and let the mixture rise. When the sponge is light, add the beaten egg, the fat and the brown sugar creamed, and the remaining flour. Knead until the dough is smooth and allow it to rise until it is double in bulk. Then roll the dough until it is 1/2 inch thick, place it in a shallow pan, and let it rise until it is light. Brush the top with 1 tablespoonful of melted butter and sprinkle it with 3 teaspoonfuls of cinnamon and 3 tablespoonfuls of sugar. Bake 10 to 15 minutes in a moderately hot oven.

CINNAMON ROLLS.—To make cinnamon rolls, which are preferred by some persons to coffee cake, use may be made of the preceding coffee-cake recipe. However, instead of rolling the dough 1/2 inch thick, roll it 1/4 inch thick and brush it with melted butter. Then sprinkle it with 1 tablespoonful of cinnamon, 1/2 cupful of light-brown sugar, and 1/2 cupful of chopped raisins. Next, roll this as a jelly roll and cut the roll into 1/2-inch slices, as shown in Fig. 22. Place these slices close together in a shallow pan and let them rise until they are light, as in Fig. 23. Then bake them in a hot oven for about 15 minutes.

## Toast

As every one knows, TOAST is sliced bread browned by means of heat. To make toast is not a difficult process, but a certain amount of care must be exercised if good results are desired. The slices used for toast may be cut thick or thin, depending on whether the persons for whom the toast is made prefer a soft or a dry toast and whether the digestibility of the toast is to be taken into consideration. If thick slices are used and they are toasted the usual length of time necessary to make the surfaces brown, the centre of the slices will remain soft. Toast made of thin slices and toasted over a slow fire becomes dry and crisp during the process of browning and is more digestible than that which is moist. Such toast will not lose its crispness unless the pieces are piled in a heap while they are hot and are allowed to soften from the moisture that collects. While toast is usually served in the form of slices, just as they are cut from the loaf, the pieces may be cut into shapes of various kinds; in fact, toast becomes more attractive if it is cut in unusual shapes. The crust of toast may be trimmed off or left on, as desired.

If the best results are desired in the making of toast, considerable attention must be given to the heat that is to produce the toast. Whatever kind is employed, it should be steady and without flame. Before a coal or a coke fire is used for this purpose, it should be allowed to burn down until the flame is gone and the coals are hot enough to reflect the heat for toasting. If a gas toaster is used, the gas should be turned sufficiently low for the bread to brown slowly. Very good results are obtained from the use of an electric toaster, also. This device has become a rather common household article where electricity is used in the home, and by means of it the toast can be made on the table and served while it is fresh and hot. In whatever way toast is made, it will lose much of its attractiveness unless it is served while it is fresh and before it loses its heat. If toast becomes burned, either from a flame that is too hot or from inattention on the part of the person who is preparing it, it may be made fit for use by scraping it lightly with a knife or by rubbing it across a grater, so as to remove the burned portion.

MILK TOAST.—Milk and toast make a combination that is liked by many persons, and when these two foods are combined the result is known as milk toast. To make milk toast, simply pour over the toast rich milk that has been heated and seasoned with salt, a little sugar, and a little butter. Thin white sauce may also be used for this purpose if desired.

FRENCH TOAST.—Possibly no dish in which toast is used is better known than the so-called French toast. Both milk and egg are used in making this dish, and these of course add to the food value of the bread. French toast made according to the following recipe will prove very satisfactory.

**French Toast**

(*Sufficient to Serve Eight*)

1 egg

1 c. milk

2 tsp. sugar 8 slices of bread 1/2 tsp. salt

Beat the egg and add it to the milk, salt, and sugar. Dip each slice of bread into this liquid, turn it quickly, and then remove it. Place the bread thus dipped in a hot frying pan and sauté it until the under side is brown; then turn it and brown the other side. Serve hot with sirup or jelly.

**Left-over Bread**

Bread that has become stale need not be wasted, for there are many uses to which it may be put. As such bread has lost much of its moisture, it is desirable for toast, for it browns more quickly and makes crisper toast than fresh bread. Thick slices of it may also be cut into cubes or long, narrow strips and then toasted on all sides, to be served with soup instead of crackers. Still another use that can be made of stale bread is to toast it and then cut it into triangular pieces to be served with creamed dishes or used as a garnish for meats, eggs, and various entrées. Left-over toast may also be cut in this way and used for these purposes.

The ends of loaves, crusts trimmed from bread used for sandwiches, or stale bread or rolls that cannot be used for the purposes that have been mentioned can also be utilised, so none of them need be thrown away. If such pieces are saved and allowed to dry thoroughly in the warming oven or in an oven that is not very hot, they may be broken into crumbs by putting them through a food chopper or rolling them with a rolling pin. After the crumbs are obtained, they should be put through a coarse sieve in order to separate the coarse ones from the fine ones. Such crumbs, both coarse and fine, may be kept for some time if they are put into jars or cans.

It is a very good plan to keep a supply of bread crumbs on hand, for there are numerous dishes that require the use of bread in this form. For instance, bread crumbs are used for all kinds of scalloped dishes; for making puddings, such as bread pudding, brown Betty, etc.; for stuffing fish, fowl, and such vegetables as tomatoes and peppers; for covering the top of baked dishes, such as various egg and cheese dishes; for breading steaks and chops; and for covering croquettes or oysters that are to be fried. They may also be added to muffins, griddle cakes, and even yeast-bread dough. With so many uses to which bread crumbs can be put, no housewife need be at a loss to know how to utilise any scraps of bread that are not, for some reason, suitable for the table.

# 6

# Hot Breads

## REQUIREMENTS AND PROCESSES FOR MAKING HOT BREADS

### Hot Breads in the Diet

Closely related to yeast breads, or those in which yeast is used as the leavening agent, are breads known as HOT BREADS, or QUICK BREADS. As these names indicate, such breads are prepared in a very short time and are intended to be served while they are fresh and hot. Hot breads, to call such breads by the name in common use, are made by baking a batter or a dough mixture formed by mixing flour, liquid, salt, and a leavening agent. The nature of the mixture, however, is governed by the proportion of flour and liquid, the two ingredients that form the basis of all bread mixtures; and by incorporating with them such ingredients as eggs, sugar, shortening, flavouring, fruits, nuts, etc. there may be produced an almost endless variety of appetising hot breads, which include popovers, griddle cakes, waffles, muffins, soft gingerbread, corn cake or corn bread, Boston brown bread, nut loaf, and baking-powder and beaten biscuit.

Because of the variety these hot breads afford, they help considerably to relieve the monotony of meals. In fact, the housewife has come to depend so much on breads of this kind that their use has become almost universal. As is well known, however, certain kinds are typical of certain localities; for instance, beaten biscuit and hoe cake are characteristic of the Southern States of the United States, while Boston brown bread is used most extensively in the New England States and throughout the East. The popular opinion of most persons is that hot breads are injurious. It is perhaps true that they may be injurious to individuals afflicted with some digestive disturbance, but, at any rate, the harmful effect may be reduced to a minimum by the correct preparation and baking of these foods.

## Principal Requirements for Hot Breads

Hot breads are quickly and easily made, but in this part of cookery, as in every other phase of it, certain principles must be understood and applied if the most satisfactory results are desired. These principles pertain chiefly to the ingredients used, the way in which they are measured and handled, the proportions in which they are combined, the necessary utensils, and the proper baking of the mixtures that are formed.

In the first place, the quality of the ingredients should be carefully considered, because on this depends the quality of the finished product. No one who prepares foods can expect good food to result from the use of inferior materials. Next, the proportion of the ingredients demands attention, for much importance is attached to this point. For instance, in making a certain kind of hot bread, the quantity of flour to be used is regulated by the quantity of bread that is desired, and the quantity of flour governs, in turn, the quantities of liquid, leavening, and other ingredients that are to be put into the mixture. When the proportions of ingredients required for a hot bread are known, it is necessary that the ingredients be measured very accurately. Leavening material, for example, will serve to make clear the need for accuracy in measuring. A definite quantity of leavening will do only a definite amount of work. Therefore, if too little or too much is used, unsatisfactory results may be expected; and, as with this ingredient, so it is with all the materials used for hot breads.

The handling of the ingredients and the mixture has also much influence on the success with which hot breads are produced. A heavy touch and excessive handling, both of which are usually characteristic of the beginner, are more likely to result in a tough product than is the light, careful handling of the expert. However, as skill in this matter comes with practice, no discouragement need result if successful results are not forthcoming at the very start in this work. A good rule to follow in this particular, and one that has few exceptions, is to handle and stir the ingredients only enough to blend them properly.

In addition to the matters just mentioned, the utensils in which to combine the hot-bread materials and bake the batters or doughs are of importance. While none of these is complicated, each must be of the right kind if the best results are expected. The final point to which attention must be given is the baking of this food. Proper baking requires on the part of the housewife familiarity with the oven that is to be used, accuracy in judging temperature, and a knowledge of the principles underlying the process of baking.

## LEAVENING AGENTS

### Classes of Leavening Agents

As has been pointed out, the ingredients that are actually required in the making of hot breads are flour, liquid, salt, and leavening, and to give variety to breads of this kind, numerous other materials, including sugar, shortening, eggs, fruit, nuts, etc., are often added. With the exception of leavening agents, none of these ingredients requires special attention at present; however, the instruction that is given in *Bread* regarding flour should be kept in mind, as should also the fact that all the materials for hot breads should be of the best quality that can be obtained.

As is known by this time, leavening agents are the materials used to leaven, or make light, any kind of flour mixture. These agents are of three classes, namely, *organic, physical,* and *chemical.* The organic agent is the oldest recognized leavening material, it being the one that is used in the making of yeast breads; but as a complete discussion of this class of leavening agents is given in *Bread* and as it is not employed in the making of hot breads, no consideration need be given to it here. Physical leavening is accomplished by the incorporation of air into a mixture or by the expansion of the water into steam, and chemical leavening agents are the most modern and accurate of all the agents that have been devised for the quick rising of flour mixtures.

### Physical Leavening

PHYSICAL LEAVENING consists in aerating, or incorporating gas or air into, a mixture that is to be baked, and it is based on the principle that air or gas expands, or increases in volume, when heated. It is definitely known that when air is incorporated into dough and then heated, the air increases 1/273 of its own volume for each degree that the temperature is increased. For instance, if the temperature of an aerated mixture is 65 degrees Fahrenheit when it is put into the oven, the air or gas will have doubled in volume by the time it has reached 338 degrees Fahrenheit. Thus, the success of aerated bread depends to some extent on the temperature of the mixture when it goes into the oven. The colder it is at that time, the greater is the number of degrees it will have to rise before it is sufficiently baked, and the more opportunity will the gas have to expand.

The air or gas required for physical leavening is incorporated into a mixture by beating or folding the batter or dough itself, or by folding beaten egg whites into it. If the mixture is thin enough, the beating may be done with a spoon or an egg beater; but if it is thick enough to be handled on a board, air may be incorporated into it by rolling and folding it repeatedly. If eggs are to be used for aerating the batter or dough, the entire egg may be beaten and then added, but as more air can be incorporated into the egg whites, the yolks and whites are usually beaten separately. To make the white of eggs most satisfactory for this purpose, it should be beaten stiff enough to stand up well, but not until it becomes dry and begins to break up. In adding the beaten egg white, it should be folded carefully and lightly into the mixture after all the other ingredients have been combined. Beaten egg white may be used to lighten any mixture that is soft enough to permit it to be folded in.

To insure the best results from mixtures that are to be made light by means of physical leavening agents, certain precautions must be taken. Such mixtures should be baked as soon as possible after the mixing is done, so that the gas or air will not pass out before the dough is baked. Likewise, they should be handled as lightly and quickly as possible, for a heavy touch and too much handling are often the cause of imperfect results. For baking aerated mixtures, heavy irons are better than tin muffin pans; also, the pans that are used should be heated before the mixture is put into them, so that the batter or dough will begin to expand immediately. Gem irons should be filled level with an aerated mixture.

## Chemical Leavening

CHEMICAL LEAVENING is brought about by the action of gas produced by an acid and an alkali. All chemical leavening agents are Similar in their action, and they are composed of an acid and an alkali. When an acid and an alkali are brought together in the presence of moisture and heat, the result is the rapid production of carbon dioxide, a gas that expands on being heated, just as all other gases do. In expanding, the gas pushes up the batters or doughs, and these, when baked, set, or harden, into porous shapes. In addition to forming the gas, the acid and the alkali produce a salt that remains in the bread, and it is this salt that is responsible for the harmful effect usually attributed to chemical leavening agents.

The first chemical leavening agents were devised by housewives themselves. They consisted of a combination of saleratus, an alkali made from wood ashes, and sour milk or molasses. The results obtained were more or less satisfactory, but never entirely accurate or certain.

Later on, chemists by employing the same idea combined an alkali with an acid in powder form and produced an accurate and satisfactory leavening agent in the form of baking powder. The discovery of baking powder, however, has not displaced the use of other combinations that form chemical leavening agents, for soda is still combined with sour milk, molasses, and cream of tartar in the making of various hot breads. Therefore, so that a proper understanding of the various chemical leavening agents may be obtained, a discussion of each is here given.

## Soda and Sour Milk

When soda is used with sour milk for leavening purposes, the lactic acid in the milk is so acted upon by the soda as to produce gas. However, these two ingredients—soda and sour milk—do not make an absolutely accurate leavening agent, because the quantity of acid in the sour milk varies according to the fermentation that has taken place. For example, sour milk 48 hours old contains more acid than sour milk that is kept under the same conditions but is only 24 hours old.

The proportion of these ingredients that is usually effective in batters and doughs for hot breads is *1 level teaspoonful of soda to 1 pint of sour milk.* So as to derive the best results in using these chemical leavening agents, it will be well to observe that if they are mixed together in a cup the milk will bubble and may, provided the quantity is sufficient, run over. These bubbles are caused by the gas that is formed when the acid and soda meet, and when they break gas escapes, with the result that some of it is lost. Formerly, it was the custom to mix these leavening substances in this way, and then to add them to the other ingredients. Now, however, in order that all gas produced may be kept in the dough mixture, the soda is sifted in with the dry ingredients and the sour milk is added with the liquid ingredients.

A point well worth remembering is that sour milk and soda may be substituted for sweet milk and baking powder in a recipe that calls for these ingredients by using *1 teaspoonful of soda to each pint of sour milk.* This information should prove valuable to the housewife, especially if she has accumulated a supply of sour milk that should not be wasted. Occasionally it will be found that baking powder and soda are required in the same recipe, but this occurs only when an insufficient amount of soda to produce the desired result is specified.

## Soda and Molasses

Although molasses, which is a product of sugar cane, is sweet, it contains an acid that is formed by the fermentation that continually occurs in it, an evidence of which is the tiny bubbles that may be seen in molasses, especially when it is kept in a warm place. Because of the presence of this acid, molasses may be used with soda to form a chemical leavening agent, and when they are combined in hot breads or cake, the chemical action of the two produces carbon dioxide.

However, accurate results cannot always be obtained when these ingredients are used, for the degree of acidity in molasses is as uncertain as it is in sour milk. Molasses that is old or has been kept in a warm place will contain more acid than molasses that has been manufactured only a short time or that has been kept cool to retard fermentation.

The proportion of soda to molasses that can usually be relied on for hot breads and cakes is *1 teaspoonful of soda to 1 cupful of molasses,* or just twice the quantity of soda that is generally used with sour milk. To produce the best results, the molasses should be mixed with the liquid ingredients and the soda sifted in with the dry ones. As molasses burns very quickly in a hot oven, all breads or cakes containing it as an ingredient should be baked in an oven of moderate temperature.

## Soda and Cream of Tartar

Some housewives are inclined to use soda and cream of tartar for leavening purposes; but there is really no advantage in doing this when baking powder can be obtained, for some baking powders are a combination of these two ingredients and produce the same result. In fact, the housewife cannot measure soda and cream of tartar so accurately as the chemist can combine them in the manufacture of baking powder. Nevertheless, if their use is preferred, they should be measured in the proportion of *twice as much cream of tartar as soda.* As in the case of soda alone, these leavening agents should be sifted with the dry ingredients. A small quantity of cream of tartar is used without soda in such mixtures as angel-food cake, in which egg white alone is used to make the mixture light. The addition of the cream of tartar has the effect of so solidifying the egg white that it holds up until the heat of the oven hardens it permanently.

## Baking Powder

Without doubt, baking powder is the most satisfactory of the chemical leavening agents. It comes in three varieties, but they are all similar in composition, for each contains an alkali in the form of soda and an acid of some kind, as well as a filler of starch, which serves to prevent the acid and the alkali from acting upon each other. When moisture is added to baking powder, chemical action sets in, but it is not very rapid, as is apparent when a cake or a muffin mixture is allowed to stand before baking. The bubbles of gas that form in such a mixture can easily be observed if the mixture is stirred after it has stood for a short time. When both moisture and heat are applied to baking powder, however, the chemical action that takes place is more rapid, and this accounts for its usefulness in baking hot breads and cake.

The price of the different kinds of baking powder, which usually varies from 10 cents to 50 cents a pound, is generally an indication of the ingredients that they contain. Powders that sell for 40 to 50 cents a pound usually contain cream of tartar for the acid, the high price of this substance accounting for the price of the powder. Powders that may be purchased for 30 to 40 cents a pound generally contain acid phosphate of lime, and as this substance is cheaper than cream of tartar, a baking-powder mixture containing it may well be sold for less. The cheapest grade of powders, or those which sell for 10 to 25 cents a pound, have for their acid a salt of aluminum called alum. Still other powders that are sometimes made up to sell for 20 to 30 cents a pound contain a mixture of phosphate and alum.

As baking powders vary in price, so do they vary in their keeping qualities, their effectiveness, and their tendency toward being injurious. Most phosphate and alum powders do not keep so well as the cream-of-tartar powders, and the longer they are kept, the less effective do they become.

The powders that contain phosphate yield more gas for each teaspoonful used than do the other varieties. Much controversy has taken place with regard to the different kinds of baking powder and their effects on the digestive tract, but authorities have not yet agreed on this matter. However, if foods made with the aid of baking powders are not used excessively, no concern need be felt as to their injurious effect.

The housewife in her choice of baking powder should be guided by the price she can afford to pay and the results she is able to get after she has become well informed as to the effect of the different varieties. She may easily become familiar with the composition of baking powder, for a statement of what substances each kind contains is generally found on the label of every variety. This information is invaluable to the housewife, as it will assist her considerably in making a selection.

The proportion of baking powder to be used in a batter or a dough is regulated by the quantity of flour employed and not. as is the case with soda and molasses or sour milk, by the quantity of liquid, the usual proportion being *2 level teaspoonfuls to 1 cupful of flour*. Sometimes this proportion is decreased, 6 or 7 teaspoonfuls being used instead of 8 to each quart of flour in the making of large quantities of some kinds of baked foods. In adding baking powder to a mixture, as in adding other dry leavening agents, it should be sifted with flour and the other dry ingredients.

Although baking powder may be purchased at various prices, a good grade can be made in the home without much effort and usually for less than that which can be bought ready made. For these reasons, many housewives prefer to make their own. The following recipe tells how to make a cream-of-tartar powder that is very satisfactory:

## Recipe for Baking Powder

1/2 lb. cream of tartar 1/4 lb. bicarbonate of soda 1/4 lb. corn starch

Weigh all the ingredients accurately. If the cream of tartar and the bicarbonate of soda are to be purchased from a druggist, it will be better for him to weigh them than for the housewife, as he uses scales that weigh accurately. After all the ingredients are weighed, mix them together thoroughly by sifting them a number of times or by shaking them well in a can or a jar on which the lid has been tightly closed. The baking powder thus made should be kept in a can or a jar that may be rendered air-tight by means of a lid, or cover.

# HOT-BREAD UTENSILS AND THEIR USE

## Purpose of Utensils

The utensils required for the making of hot breads consist of two kinds: those in which the ingredients are prepared and combined to form the mixture and those in which the mixture is to be baked. As soon as it is known just what ones are needed to carry out the recipe for the hot bread that is to be made, they, together with the necessary ingredients, such as milk, fat, flour, baking powder, salt, eggs, etc., should be collected and arranged in the manner shown in Fig. 1, so that they will be convenient. Usually, much of the success of hot breads depends on the quickness and dexterity with which the ingredients are put together, and if the person making them has to interrupt her work every now and then to get out a utensil, she will find that her results will not be so satisfactory and that she will use up more energy than the work really demands. The pans in which the mixture is to be baked need particular attention, for they should be greased and ready to fill before the mixing is begun. If they are to be heated, they should be greased and put into the oven a few minutes before the mixture is ready to be put into them, so that they may be taken from the oven and filled at once.

## Utensils for Preparing the Mixture

These consist of a bowl *a* of the proper size for mixing; a smaller bowl *b* for beating eggs, provided eggs are to be used; two standard half-pint measuring cups *c*, one for dry ingredients and the other for wet ingredients; a tablespoon *d*, a case knife *e*, and a teaspoon *f* for measuring and mixing; an egg beater *g* and a flour sifter. Of course, if an egg whip is preferred, it may take the place of the egg beater, but for some hot-bread mixtures use will be found for both of these utensils.

## Utensils for Baking the Mixture

The kind of utensil required for the baking of hot-bread mixtures depends entirely on the nature of the mixture and the recipe that is to be prepared. Muffins require muffin pans like those illustrated at *h;* Boston brown breads need cans that have tight-fitting lids; soft ginger bread, nut loaf, and corn cake are baked in loaf pans; baking-powder or beaten biscuits are placed in shallow pans or on oiled sheets; griddle cakes must be baked on griddles; and waffles require waffle irons. None of these utensils are likely to present any difficulty in their use except griddles and waffle irons, so in order that these may be thoroughly understood and good results thereby obtained, explanations of them are here given.

## Griddles

Such utensils are made of numerous materials, but the most satisfactory ones are constructed of steel, iron, soapstone, and aluminum. Steel and iron griddles must be greased before cakes are baked on them so as to prevent the cakes from sticking; for this reason they are less convenient than soapstone and aluminum griddles, which do not require any grease.

The size of griddle to use is governed by the number of persons that are to be served. One that is unusually large, however, should be avoided if a gas stove is used for cooking, as it is difficult to heat a large griddle evenly on such a stove, and even a small one must be shifted frequently so that some spots will not be hotter than others. In this respect, a griddle made of aluminum has the advantage over the other kinds, for this material conducts the heat evenly over its entire surface.

Before a new steel or iron griddle is used, it must be tempered so as to prevent the food that is to be baked on it from sticking. If it is not tempered, much time will be consumed before its surface will be in the right condition to permit baking to proceed without difficulty, and this, of course, will result in wasting considerable food material. Tempering may be done by covering the griddle with a quantity of fat, placing it over a flame or in a very hot oven, and then allowing it to heat thoroughly to such a temperature that the fat will burn onto the surface. This same precaution should be observed with new waffle irons and frying pans made of steel or iron if the best results from such utensils are desired.

WAFFLE IRONS.—A waffle iron, as shown in Figs. 4 and 5, consists of two corrugated griddles fastened together with a hinge in such a way that the surfaces nearly touch when the handles are brought together as in Fig. 4 (*a*). These griddles are so suspended in a frame that they may be turned completely over in order to allow each side to be exposed to the heat. The waffle iron illustrated in Fig. 4, shown closed in view (*a*) and open in (*b*), is intended for a coal range. In order to use it, a stove lid is removed from one of the openings and the waffle iron is set in the opening, which allows the griddle part to be turned. The waffle iron shown in Fig. 5 is intended for a gas range. As will be noticed, the griddle part rests on a base that is deep enough to permit it to be turned. In using a waffle iron of either kind, it should be heated while the waffle mixture is being prepared; then it should be thoroughly greased on both sides. No excess fat, however, should be used, as it will run out when the griddle is turned over.

## THE MIXTURE

### Varieties of Mixtures and General Proportions

The mixtures from which hot breads are produced are of different consistencies, and familiarity with them is necessary if good results in the making of such breads are desired. This difference in the consistencies is due to the proportion of flour and liquid used, a small proportion of flour producing a *batter* and a large proportion, a *dough*. It will be well to note, however, that some kinds of flour thicken a mixture much more readily than do others. Experience in the handling of flour teaches how to vary the other ingredients of a recipe in order to make them correspond to the difference in flour, but the person who lacks a knowledge of cookery, or has had very little experience in the handling of foods, must know the general proportions that are correct under most circumstances. The names of the mixtures that the ingredients produce are *thin batter, thick batter, soft dough,* and *stiff dough.*

A THIN BATTER is one in which the general proportion of liquid and flour is *1 measure of flour* to *1 measure of liquid*. Such a batter, when poured, immediately seeks its own level and has the consistency of thin cream. The most common examples of thin batters are popovers and griddle cakes.

A THICK BATTER, which is known as a *drop*, or *muffin, batter*, is one that is made of *2 measures of flour* and *1 measure of liquid*. A batter of this kind may be poured, but it will not immediately seek its own level. Muffins, gems, puddings, and cakes are made of thick batters.

A SOFT DOUGH is one whose proportions are *3 measures of flour* and *1 measure of liquid*. A dough of this kind will stand up alone—that is, without support at the sides—and has more of the properties of a solid than of a liquid. Baking-powder biscuits, tea rolls, and certain kinds of cake are made of this form of dough.

A STIFF DOUGH is made of *4 measures of flour* and *1 measure of liquid*. Such a dough will not cling to the mixing bowl, can be handled with the hands, and will not stick when rolled out on a board. Pie crust, hard cookies, and beaten biscuit are made of such dough.

APPLYING KNOWLEDGE OF GENERAL PROPORTIONS. While the general proportions just mentioned remain the same in the majority of cases, they vary somewhat when ingredients other than liquid and flour are added. Shortening and eggs in particular change the quantity of liquid required, less liquid being necessary when these ingredients are used. To get the best results from a new recipe, it is always advisable upon reading the recipe to notice the proportions that are given and then to try to judge whether they bear a close enough resemblance to the general proportions to make a successful dish. For instance, if a griddle-cake recipe calls for 3 cupfuls of flour and 1 cupful of liquid, the cook who understands what the general proportions for such a batter ought to be would know immediately that the recipe calls for too much flour. Likewise, she would know that a recipe for baking-powder biscuits that calls for 2 cupfuls of flour and 1 cupful of liquid would make a dough that would be too soft to handle. Besides enabling a woman to judge a recipe, a knowledge of the correct proportions for things of this kind makes it possible for her to combine the ingredients for a certain recipe without resorting to a cook book, or, in other words, to originate a recipe. Because of the importance of such an understanding, attention should always be given to details that will assist in obtaining a thorough knowledge of this matter.

## Preparing the Mixture

PRELIMINARY PREPARATION OF INGREDIENTS.—Before the mixing of the ingredients that are to be used in the batters and doughs of hot breads is begun, all that are needed for the recipe selected should be collected and properly measured. Always sift the flour that is to be used for this purpose. This is a rule that never varies with regard to flour to be used for any dough mixture or as a thickening agent. Then, to prevent the flour from packing too solidly, measure it by dipping it into the cup with a spoon. To obtain the proper amount, heap the cup and then level it with the edge of a knife. Measure with a spoon whatever dry leavening agent is called for, and be sure that it does not contain any lumps. If salt, sugar, and spices are to be used, measure them carefully. Mix the leavening agent, the salt, the sugar, and the other dry ingredients with the flour by sifting them together once or twice. Measure the butter or other fat by packing it in the spoon and then leveling it with a knife. Be particular in measuring the liquid, using neither more nor less than is called for. Regarding this ingredient, it should always be remembered that when a cupful is required, a half-pint cup full to the brim is meant and that any fraction of a cupful should be measured with the same exactness.

COMBINING THE INGREDIENTS.—The manner in which a batter or a dough is mixed is very important, for much of the success of the finished product depends on the order in which the various steps are accomplished. Two general methods of combining the ingredients for such mixtures have been devised and either of them may be followed, because they produce equally good results.

In one of these methods, the fat is worked into the dry ingredients and the liquid then added. As eggs are usually considered a liquid ingredient, they are beaten and added to the rest of the liquid before it is mixed with the dry ingredients. However, if eggs are to be used for leavening, only the yolks are added with the liquid ingredients, the whites being beaten separately and folded in last.

The other method is used only when the mixtures are to contain a small quantity of fat. In this method, all the liquid ingredients, including the eggs, are first mixed together. Then the dry ingredients are combined and sifted into the liquid. The fat is melted last and beaten into the dough mixture. If the mixture to be handled is a stiff one, the fat should be put in cold, for adding melted fat makes the dough soft and sticky and therefore difficult to handle.

## Baking the Mixture

REGULATING THE OVEN.—When the ingredients have been properly combined, the mixture is ready to be baked. With the exception of waffles and griddle cakes, the baking of which is explained in connection with the recipes, all hot breads are baked in the oven; therefore, while the mixture is being prepared, the oven should be properly regulated in order that the temperature will be just right when it is time to start the baking. Particular thought should be given to this matter, for if no attention is paid to the oven until the mixture is ready to be baked, it will be necessary to allow the mixture to stand until the heat of the oven can be regulated or to put it into the oven and run the risk of spoiling the food. To prevent either of these conditions and to insure success, the fuel, no matter what kind is used, should be lighted before mixing is begun, so that the oven may be heating while the mixture is being prepared, unless, as is sometimes the case, there are steps in the preparation of the mixture that consume considerable time. For instance, looking over raisins and cleaning them or cracking nuts and picking the meats out of the shells should be done before the rest of the ingredients are prepared or the oven is regulated.

CORRECT OVEN TEMPERATURES.—Quick breads that are to be baked in the form of loaves require an oven temperature of from 350 to 400 degrees Fahrenheit. Muffins, biscuits, and the smaller varieties of these breads need a higher temperature, 425 to 450 degrees Fahrenheit being best. As they are not so large, the heat has less dough through which to penetrate, and consequently the baking can be accomplished more quickly.

DETERMINING AND REGULATING OVEN TEMPERATURE.— Regulating the oven and testing its temperature present very little difficulty to the housewife of experience, but they are not always easy problems for the woman who is learning to cook. However, if the untrained and inexperienced cook will observe her oven closely and determine the results of certain temperatures, she will soon find herself becoming more successful in this matter. To assist the housewife in this matter, as well as to help in the saving of much loss in fuel and in underdone or overdone food, many stoves are equipped with an oven thermometer, an indicator, or a thermostat. The thermometer is more likely to be reliable than the indicator, as it has a column of mercury like that of any other thermometer and is graduated; also, a certain kind may be secured that can be used with any sort of oven. The indicator is in the form of a dial with a hand attached to a metal spring. This spring contracts and expands with the changes in the temperature of the oven and thus causes the hand to point out the temperature. The thermostat is a device that automatically regulates the heat of the oven. On a stove equipped with a thermostat, it is simply necessary to set the device at the temperature desired. When this temperature is reached, the device keeps it stationary.

If neither an indicator nor a thermometer is available, the heat of the oven may be determined in other ways. Some housewives test the oven with the hand, and while such a test is more or less dependent on experience, those who use it find it very satisfactory. If the hand can be held in the oven while 15 is counted slowly, the temperature is that of a moderate oven and will be right for the baking of loaves. An oven that is of the proper temperature for muffins or rolls will permit the hand to be held in it while only 10 is counted slowly. Those who do not test with the hand find that placing a piece of white paper in the oven is an accurate way of determining its temperature. Such paper will turn a delicate brown in 5 minutes in a moderate oven, and a deeper brown in 4 minutes in a hot oven.

PROPER PLACING OF THE MIXTURE IN THE OVEN.—As is pointed out in *Essentials of Cookery,* Part 1, the top of the oven is hotter than the bottom. This truth and the fact that in an oven, as in any other space, air expands and rises on becoming heated, are points that have much to do with the baking of quick breads, for these are mixtures that rise after being placed in the oven. So that they may rise properly, they should be placed on the bottom first; then, as they become heated, they will have a tendency to rise as the air does. If the food is placed near the top first, the heated air will be likely to press it down and retard its rising. As soon as the rising is completed and the food has baked sufficiently on the bottom, it should be moved up so that it will brown on the top.

TESTING THE BAKED MIXTURE.—Recipes for baked dishes usually state the length of time required to bake them, but such directions cannot always be depended on, because the temperature of the oven varies at different times. The best way in which to judge whether the food has baked the necessary length of time is to apply to it one of the reliable tests that have been devised for this purpose.

Probably the most satisfactory test is to insert a toothpick as deep as possible into the center of the loaf. The center, rather than some other part of the loaf, is the place where the testing should be done, because the heat penetrates a mixture from the outside and the center is therefore the last part to bake. If the toothpick comes out without particles of dough adhering, the mixture is sufficiently baked in that place and consequently throughout the loaf. In case the dough sticks to the toothpick, the baking is not completed and will have to be continued. Since this is a test that is frequently used, a supply of toothpicks, preferably round ones, should be kept in a handy place near the stove.

Another fairly accurate means of testing baked mixtures that do not form a very hard crust consists in making a dent in the center with the finger. If the dent remains, the baking must be continued, but if it springs back into place, the baking is completed.

## Serving Hot Breads

Hot breads, in contrast with yeast breads, are intended to be eaten hot, and, to be most satisfactory, should be served as soon as possible after they are baked. They usually take the place of bread in the meal for which they are served, but there are various ways of using them whereby variety is given to them and to the meal. A favorite combination with many persons is hot biscuits or muffins served with honey. If honey is not available, jam, preserves, or sirup may be substituted to advantage. A mixture made like baking-powder biscuits and baked or steamed is especially good when served with chicken or meat stew poured over it. The same mixture sweetened and made a trifle richer may be served with fruit and cream for short cake. For afternoon tea, tiny muffins and biscuits about the size of a 50-cent piece are very attractive. Then, too, if they are split and buttered, they may be served with salad for a light luncheon.

Hot breads baked in the form of a loaf require some attention as far as preparing them for the table is concerned. Gingerbread and corn cake are better if they are broken rather than cut while hot. In case they are preferred cut, a sharp knife should be employed, and, to obtain slices that have a good appearance, the knife should be heated and the cutting done before it cools. Usually, gingerbread is served plain, but the addition of icing improves it considerably and provides a simple cake that can be used for dessert.

RECIPES FOR HOT BREADS

## Popover Recipes

POPOVERS.—A delightful change from the puffs, muffins, and biscuits that are usually served for breakfast or luncheon is afforded by means of popovers, one of which is illustrated in Fig. 6. Popovers are not difficult to make. For them is required a thin batter in equal proportions of liquid and flour. In giving the method for mixing popovers, some of the older cook books recommend beating for 5 minutes just before they are baked, because the lightness was formerly supposed to be due to the air that is incorporated by this beating. It is possible, however, to make very light popovers with only enough beating to mix the ingredients thoroughly, and it is now known that the rising is due to the expansion of water into steam in the mixture. This knowledge is useful in that it saves time and energy.

## Popovers

(*Sufficient to Serve Six*)

1 c. flour 1/4 tsp. salt 1 c. milk

1 egg

Mix the flour, salt, and milk in a bowl, and then drop in the unbeaten egg. Beat all with a rotary egg beater until the mixture is perfectly smooth and free from lumps. Grease and warm gem irons or popover cups. Then fill them about two-thirds full of the popover batter. Bake in a moderate oven for about 45 minutes or until the popovers can be lifted from the cups and do not shrink when removed from the oven.

POPOVERS WITH FRUIT.—Popovers made according to the preceding recipe are particularly good if fruit is added to them. To add the fruit, cut a slit in the side of the popovers as soon as they are removed from the oven and insert a few spoonfuls of apple sauce, marmalade, preserves, jelly, or canned fruit. These may be served either warm or cold as a breakfast dish, or they may be sprinkled with powdered sugar and served with cream for a dessert or a luncheon dish.

NUT PUFFS.—An example of a thin batter not in equal proportions of liquid and flour is afforded by nut puffs. In hot breads of this kind, aeration is used as the leavening agent. In order to assist with the incorporation of air, the egg yolk is well beaten before it is added; but the greater part of the lightness that is produced is due to the egg white, which is beaten and folded in last. The addition of nuts to a batter of this kind considerably increases its food value.

**Nut Puffs**

(*Sufficient to Serve Six*)

1-1/2 c. flour

2 Tb. sugar

1 tsp. salt

1 c. milk

1 egg

1 Tb. fat

1/4 c. chopped nuts

Sift the flour, sugar, and salt together, and add the milk and beaten egg yolk. Melt the fat and add it and the chopped nuts. Beat the egg white stiff and fold it into the mixture carefully. Fill hot, well-greased gem irons level full of the batter, and bake in a hot oven about 20 minutes.

WHOLE-WHEAT PUFFS.—Puffs in which use is made of whole-wheat flour instead of white flour are also an example of a thin batter that is made light by aeration. If desired, graham flour may be substituted for the whole-wheat flour, but if it is a coarser bread will be the result. This coarseness, however, does not refer to the texture of the bread, but is due to the quantity of bran in graham flour. Whole-wheat puffs, as shown in Fig. 7, are attractive, and besides they possess the valuable food substances contained in whole-wheat flour, eggs, and milk.

## Whole-wheat Puffs

(*Sufficient to Serve Six*)

1-1/2 c. whole-wheat flour

2 Tb. sugar

1 tsp. salt

1 c. milk

1 egg

1 Tb. fat

Sift the flour, sugar, and salt together and add the milk and the egg yolk, which should be well beaten. Melt the fat and stir it into the batter. Beat the egg white stiff, and fold it in carefully. Heat well-greased gem irons, fill them level full with the mixture, and bake in a hot oven for about 20 minutes.

## Griddle-cake Recipes

PROCEDURE IN BAKING GRIDDLE CAKES.—During the preparation of the batter for griddle cakes, have the griddle heating, so that it will be sufficiently hot when the cakes are ready to be baked. Each time, before the baking is begun, grease the griddle, provided it is the kind that requires greasing, by rubbing over it a rind of salt pork or a small cloth pad that has been dipped into a dish of grease. In greasing the griddle, see that there is no excess of grease, as this burns and produces smoke.

When the griddle has become hot enough for the batter to sizzle when it is put on, the baking may be started. Pour the batter on the griddle from the tip of a large spoon, so that the cakes will form as nearly round as possible. When the top surface is full of bubbles, turn the cakes with a spatula or a pancake turner, and allow them to brown on the other side. By the time the cakes are sufficiently browned on both sides, they should be cooked through and ready to serve. If they brown before they have had time to cook through, the griddle is too hot and should be cooled by moving it to a cooler part of the stove or by reducing the heat. A very important point to remember in the baking of griddle cakes is that they should not be turned twice, as this has a tendency to make them heavy.

GRIDDLE CAKES.—As is generally known, griddle cakes are thin batters that are made light with a chemical leavening agent. Eggs are often used in such batters, but it is possible to make very excellent griddle cakes without the use of any eggs. It should also be remembered that the use of too much egg is more certain to make the cakes tough and less palatable than if none is used. The kind of flour used for griddle cakes has much to do with the consistency of the batter used for them.

If, when the first cakes are placed upon the griddle, the batter seems to be either too thick or too thin, liquid or flour may be added to dilute or thicken the batter until it is of the right consistency. For instance, if bread flour is used, more liquid may be needed, and if pastry flour is used, more flour may be required.

## Griddle Cakes

(*Sufficient to Serve Six*)

3 c. flour

5 tsp. baking powder

1 tsp. salt

1/4 c. sugar

1 egg 2-1/4 c. milk

2 Tb. melted fat

Mix and sift the flour, baking powder, salt, and sugar. Beat the egg, add to it the milk, and pour this liquid slowly into the dry ingredients. Beat the mixture thoroughly and then add the melted fat. Bake the cakes on a hot griddle as soon as possible after the batter is mixed.

SOUR-MILK GRIDDLE CAKES.—Very delicious griddle cakes may be made by using sour milk and soda for the liquid and leavening instead of sweet milk and baking powder. Besides being particularly appetising, such cakes serve to use up left-over milk that may have soured.

There is very little difference between the ingredients for this recipe and one calling for sweet milk, except that sour milk, which is a trifle thicker in consistency than sweet milk, requires less flour to thicken the mixture.

### Sour-milk Griddle Cakes

(*Sufficient to Serve Six*)

2-1/2 c. flour 1/2 tsp. salt 2 Tb. sugar

1 tsp. soda

2 c. sour milk (not thick)

1 egg

Mix and sift the flour, salt, sugar, and soda. Add to these the sour milk and the egg well beaten. If the milk is thick, the quantity should be increased accordingly. Beat the mixture thoroughly and bake at once on a hot griddle.

CORN GRIDDLE CAKES.—The addition of corn meal to a griddle-cake mixture adds variety and food value and produces an agreeable flavor. Where corn meal is cheap, it is an economical ingredient to use in griddle cakes and other hot breads.

### Corn Griddle Cakes

(*Sufficient to Serve Six*)

1/2 c. corn meal 1-1/2 c. boiling water

2 c. milk

2 c. flour

5 tsp. baking powder 1-1/2 tsp. salt

1/4 c. sugar

1 egg

2 Tb. melted fat

Add the corn meal to the boiling water, boil 5 minutes, and turn into a bowl. Then add the milk. Next, mix and sift the flour, baking powder, salt, and sugar, and stir them into the first mixture. Beat the egg and add to the whole. Finally, stir in the melted fat. Bake on a hot griddle.

RICE GRIDDLE CAKES.—If a change in the ordinary griddle cakes that are used for breakfast is desired, rice griddle cakes should be tried. Besides lending variety, the addition of rice to a griddle-cake mixture helps to use up any left-over rice that may have been cooked for another purpose. Steamed or boiled rice used for this purpose should be broken up with a fork before it is mixed in the batter, so that the grains of rice will not stick together in chunks.

### Rice Griddle Cakes

(*Sufficient to Serve Six*)

2-1/2 c. flour

5 tsp. baking powder

1/4 c. sugar

1/2 tsp. salt

1/2 c. cold cooked rice

1 egg 1-1/2 c. milk

2 Tb. melted fat

Mix and sift the flour, baking powder, sugar, and salt. Work the rice into the dry ingredients. Add the egg, well beaten, the milk, and the melted fat. Bake on a hot griddle.

BUCKWHEAT CAKES.—Buckwheat flour is used for griddle cakes more than for any other purpose.

When used in this way it has a very typical flavor that most people find very agreeable. Many prepared buckwheat flours, to which have been added the quantity of leavening agent necessary to raise the mixture, are on the market for the convenience of those who do not desire to prepare the mixture at home. As a rule, these contain a combination of buckwheat and wheat flour. To make cakes from these flours, add the required amount of liquid, either milk or water, and a little sugar, if necessary, and then proceed to bake them on a griddle. While there is no objection to the use of such flours if they are found agreeable, it is more expensive to use them than to make up the buckwheat mixture at home. A recipe for buckwheat cakes that proves very satisfactory is the following:

### Buckwheat Cakes

(*Sufficient to Serve Six*)

2 c. scalded milk

1/2 c. fine bread crumbs

1/2 tsp. salt

1/4 yeast cake

3/4 c. lukewarm water

1-1/2 c. buckwheat flour

1/2 c. white flour

1 Tb. molasses

1/4 tsp. soda

Pour the scalded milk over the bread crumbs and add the salt. Dissolve the yeast cake in 1/2 cupful of the lukewarm water and add this to the bread crumbs and milk. Stir in the buckwheat and the white flour, and let the mixture rise overnight. In the morning, stir it well and add the molasses, the soda, and 1/4 cupful of lukewarm water. Bake on a hot griddle.

If cakes are to be baked the next day, retain 1/2 cupful of the batter, to which may be added flour, milk, salt, and molasses. By doing this each day, a starter may be had for a long period of time. If a strong buckwheat flavor is desired, use all buckwheat flour, but if only a slight buckwheat flavor is desired, make the proportion of wheat flour greater and that of the buckwheat smaller.

## Waffle Recipes

PROCEDURE IN BAKING WAFFLES.—The procedure in making waffles is very similar to that in making griddle cakes. While the waffle mixture is being prepared, heat the waffle iron. Then grease it thoroughly on both sides with a rind of salt pork or a cloth pad dipped in fat, being careful that there is no excess fat, as it will run out when the iron is turned over. With the iron properly greased and sufficiently hot, place several spoonfuls of the batter in the center and close the iron. By so doing, the batter will be pressed out to cover the entire surface. In pouring the batter, do not cover the entire surface of the iron with batter nor place any near the outside edge, for it is liable to run out when the iron is closed. In case this happens, be sure to put in less batter the next time. Allow the waffle to brown on the side near the fire and then turn the iron, so as to brown the other side. When the waffle is sufficiently brown, remove it; then grease the iron and repeat the process.

WAFFLES.—The form of hot bread known as waffles, which are illustrated in Fig. 8, offers the housewife an excellent opportunity to add variety to meals. Practically no one dislikes waffles, and they are especially appetising when sprinkled with powdered sugar or served with sirup. They are often served with chicken or other gravy.

## Waffles

(*Sufficient to Serve Six*)

2 c. flour

3 tsp. baking powder 1/2 tsp. salt 2 eggs 1-2/3 c. milk 2 Tb. melted fat

Sift the flour, baking powder, and salt together. Beat the yolks and whites of the eggs separately. Add the beaten yolks and the milk to the dry ingredients and then stir in the melted fat. Beat the egg whites stiff and fold them into the batter. Bake according to the directions given in Art. 45.

RICE WAFFLES.—Rice waffles offer an excellent means of utilizing left-over rice. Such waffles are prepared in about the same way as the waffles just mentioned. In working the cooked rice into the dry ingredients, use should be made of a light motion that will not crush the grains, but will separate them from one another. Left-over cereals other than rice may also be used in this way.

## Rice Waffles

(*Sufficient to Serve Six*)

1-3/4 c. flour

2 Tb. sugar

1/2 tsp. baking powder

1/2 tsp. salt

2/3 c. cooked rice

1-1/2 c. milk

1 egg

1 Tb. melted fat

Mix and sift the flour, sugar, baking powder, and salt, and then work the rice into the dry ingredients. Add the milk and the well-beaten yolk of egg. Stir in the melted fat. Beat the egg white stiff, and fold it into the batter. Bake as previously directed.

## Muffin Recipes

Muffins are examples of thick batters with variations. This form of hot bread, an illustration of which is shown in Fig. 9, may be baked in a pan like that shown at *h*, Fig. 1, or in individual tins.

Just as other forms of hot breads assist the housewife in making changes or additions to meals, so do muffins, as they are usually relished by nearly every one.

PLAIN MUFFINS.—Perhaps the simplest form of muffin is the plain, or one-egg, muffin, which is illustrated in Fig. 9 and made according to the accompanying recipe. To a plain-muffin recipe, however, may be added any kind of fruit, nuts, or other ingredients to give variety of flavour. Likewise, it may be made richer and sweeter and then steamed or baked to be served with a sauce for dessert. If it is made still richer and sweeter, the result is a simple cake mixture. Any given muffin recipe in which sweet milk is used may be made with sour milk by using soda instead of baking powder.

**Plain Muffins**

(*Sufficient to Serve Six*)

2 c. flour

2 Tb. sugar

1 tsp. salt

4 tsp. baking powder

1 c. milk

1 egg

2 Tb. melted fat

Mix and sift the flour, sugar, salt, and baking powder, and to these add the milk and beaten egg. Then stir in the melted fat. Fill well-greased muffin pans about two-thirds full of the mixture and bake in a hot oven for about 20 minutes.

BLUEBERRY MUFFINS.—Muffins containing blueberries can be made successfully only in blueberry season, but other fruit, as, for example, dates, may be used in place of the blueberries. Cranberries are often used in muffins, but to many persons they are not agreeable because of the excessive amount of acid they contain.

**Blueberry Muffins**

(*Sufficient to Serve Six*)

3 Tb. fat 1/3 c. sugar 1 egg 1 c. milk 2-1/4 c. flour 1/2 tsp. salt

4 tsp. baking powder 1 c. fresh blueberries

Cream the fat, and add the sugar gradually. Then stir in the beaten egg and milk. Reserve 1/4 cupful of flour, and mix the remainder with the salt and the baking powder. Stir the dry ingredients into the first mixture. Next, mix the 1/4 cupful of flour with the berries and fold them into the batter. Fill well-greased muffin pans about two-thirds full of the batter, and bake in a hot oven for about 20 minutes.

DATE MUFFINS.—The recipe given for blueberry muffins may be used for date muffins by substituting dates for blueberries. To prepare the dates, wash them in warm water, rinse them in cold water, and then dry them between towels. Cut them lengthwise along the seed with a sharp knife, remove the seed, and then cut each date into three or four pieces.

CORN-MEAL MUFFINS.—To many persons, corn-meal muffins, an illustration of which is shown in Fig. 10, are more agreeable than plain white-flour muffins. Corn meal gives to muffins an attractive flavour and appearance and increases their food value slightly; but perhaps its chief value lies in the variety that results from its use.

**Corn-meal Muffins**

(*Sufficient to Serve Six*)

1/2 c. corn meal

1 c. flour 3 tsp. baking powder

2 Tb. sugar 1/2 tsp. salt 3/4 c. milk

1 egg

2 Tb. melted fat

Mix and sift the corn meal, flour, baking powder, sugar, and salt. Add to these the milk and the well-beaten egg, and stir in the melted fat. Fill well-greased muffin pans two-thirds full, and bake in a hot oven for about 20 minutes.

GRAHAM MUFFINS.—A pleasing variety in the way of muffins is produced by using part graham flour,

but whole-wheat flour may be substituted for the graham flour in case it is preferred. Sour milk is used in the recipe here given, but if there is no sour milk in supply, sweet milk and baking powder may be used instead, with merely the correct proportion of soda for the molasses. If the taste of molasses is undesirable, liquid, which may be either sweet or sour milk, may be substituted for it. It is an excellent plan to be able to substitute one thing for another in recipes of this kind, and this may be done if the materials are used in correct proportion.

## Graham Muffins

(*Sufficient to Serve Six*)

1-1/4 c. graham flour 1 c. white flour 3/4 tsp. soda 1 tsp. salt 1 c. sour milk 1/3 c. molasses

1 egg

2 Tb. melted fat

Mix and sift the graham and the white flour, the soda, and the salt. Put the bran that sifts out back into the mixture. Add the milk, molasses, and well-beaten egg to the dry ingredients, and then stir in the melted fat. Fill well-greased muffin pans two-thirds full and bake in a moderate oven for about 20 minutes.

RICE MUFFINS.—Rice may be combined with white flour in the making of muffins if variety is desired.

As rice used for this purpose is added hot, it may be cooked either purposely for the muffins or for something else and only part used for the muffins. Cereals other than rice may be used in exactly the same quantity and in the same way in making muffins.

## Rice Muffins

(*Sufficient to Serve Six*)

2-1/4 c. flour

5 tsp. baking powder

2 Tb. sugar

1/2 tsp. salt

1-1/4 c. milk

1 egg 3/4 c. hot, cooked rice

2 Tb. melted fat

Mix and sift the flour, baking powder, sugar, and salt, and to these add half of the milk and the egg, well beaten. Mix the remaining half of the milk with the rice and add it to the mixture. Stir in the melted fat last. Fill well-greased muffin pans two-thirds full, and bake in a hot oven for about 20 minutes.

BRAN MUFFINS.—The particular value of bran muffins lies in the laxative quality that they introduce into the diet. In addition, they will be found to be very tasty and superior to many other kinds of muffins. Bran for such purposes as this may be bought in packages, in the same way as many cereals.

## Bran Muffins

(*Sufficient to Serve Six*)

1-1/2 c. white flour

1/2 tsp. soda

1/2 tsp. baking powder

1 tsp. salt

2 c. bran 1-1/4 c. milk 1/2 c. molasses 1 egg

Mix and sift the flour, soda, baking powder, and salt. Then add the bran, the milk, the molasses, and the well-beaten egg. Fill well-greased muffin pans about two-thirds full, and bake in a moderate oven for about 25 minutes.

## Corn-cake Recipes

CORN CAKE.—Corn cakes were among the first breads made of cereal foods in America, being at first often made of only corn meal, water, and salt.

These cakes of corn meal were prepared and carried on long journeys made by people when there were no means of rapid transportation.

The cakes did not spoil, were not bulky, and contained a great deal of nutriment, so they made a convenient kind of food for such purposes and were called *journey cakes*. From this term came the name *Johnny cake,* which is often applied to cake of this kind. The combining of flour, eggs, shortening, and sugar makes a cake that does not resemble the original very much, but in many localities such cake is still called Johnny cake.

The proportion of corn meal to flour that is used determines to a large extent the consistency of the cake; the greater the quantity of corn meal, the more the cake will crumble and break into pieces. The addition of white flour makes the particles of corn meal adhere, so that most persons consider that white flour improves the consistency.

## Corn Cake

(*Sufficient for One Medium-Sized Loaf*)

3/4 c. yellow corn meal

1-1/4 c. flour

1/4 c. sugar

3/4 tsp. salt

4 tsp. baking powder

1 c. milk

1 egg

2 Tb. melted fat

Mix and sift the corn meal, flour, sugar, salt, and baking powder. Add the milk and well-beaten egg, and then stir in the melted fat. Pour into a well-greased loaf pan and bake in a hot oven for about 30 minutes.

SOUTHERN CORN CAKE.—In the preceding recipe for corn cake, more flour than corn meal is used, but many persons prefer cake of this kind made with more corn meal than flour. Southern corn cake, which contains more corn meal and less white flour, proves very satisfactory to such persons. Therefore, which of these recipes should be used depends on the taste of those who are to eat the cake.

## Southern Corn Cake

(*Sufficient for One Medium-Sized Loaf*)

1 c. corn meal

1/2 c. flour

3 tsp. baking powder

3/4 tsp. salt

1/4 c. sugar

3/4 c. milk

1 egg

2 Tb. melted fat

Mix and sift together the corn meal, flour, baking powder, salt, and sugar. Add to them the milk and well-beaten egg, and stir in the melted fat. Pour into a well-greased loaf pan, and bake in a moderate oven for about 30 minutes.

MOLASSES CORN CAKE.—Molasses corn cake, just as its name indicates, is corn cake containing molasses.

To those who find the taste of molasses agreeable, this recipe will appeal. Others not so fond of molasses will, without doubt, prefer the plain corn cake.

Besides adding flavour, the molasses in this recipe adds food value to the product.

**Molasses Corn Cake**

(*Sufficient for One Medium-Sized Loaf*)

1 c. corn meal

3/4 c. flour

3-1/2 tsp. baking powder

1 tsp. salt

3/4 c. milk

1/4 c. molasses

1 egg

2 Tb. melted fat

Mix and sift the corn meal, flour, baking powder, and salt. Add the milk, molasses, and well-beaten egg and stir in the melted fat. Pour into a well-greased loaf pan, and bake in a moderate oven for about 30 minutes.

**Biscuit Recipes**

BAKING-POWDER BISCUITS.—The ability of the housewife as a cook is very often judged by the biscuits she makes; but they are really very simple to make, and if recipes are followed carefully and measurements are made accurately, only a little experience is required to produce excellent ones.

The principal requirement in making baking-powder biscuits, which are illustrated in Fig. 11, is that all the ingredients be kept as cold as possible during the mixing.

Tiny, thin biscuits may be split, buttered, and served with tea, while larger ones may be served with breakfast or luncheon. In order to utilise left-over biscuits of this kind, they may be split and toasted or dipped quickly into boiling water and heated in a quick oven until the surface is dry.

### Baking-powder Biscuits

(*Sufficient to Serve Six*)

2 c. flour

1 tsp. salt 4 tsp. baking powder

2 Tb. fat 3/4 c. milk

Mix and sift the flour, salt, and baking powder. Chop the fat into the dry ingredients until it is in pieces about the size of small peas. Pour the milk into the dry ingredients, and mix them just enough to take up the liquid. Make the mixture as moist as possible, and still have it in good condition to handle. Then sprinkle flour on a molding board, and lift the dough from the mixing bowl to the board. Sprinkle flour thinly over the top and pat out the dough until it is about 1 inch thick. Cut the dough with a biscuit cutter, and place the biscuits thus cut out on baking sheets or in shallow pans. If a crusty surface is desired, place the biscuits in the pan so that they are about an inch apart; but if thick, soft biscuits are preferred, place them so that the edges touch. Bake 18 to 20 minutes in a hot oven.

EMERGENCY BISCUITS.—As shown in Fig. 12, emergency biscuits resemble very closely baking-powder biscuits, and so they should, because the recipe given for baking-powder biscuits may be used for emergency biscuits by merely adding more milk—just enough to make the dough a trifle too moist to handle with the hands. When the dough is of this consistency, drop it by spoonfuls in shallow pans, as in Fig. 13, or on baking sheets. Then bake the biscuits in a hot oven for 18 to 20 minutes.

PINWHEEL BISCUITS.—To create variety, a baking-powder biscuit mixture may be made into pinwheel biscuits, a kind of hot bread that is always pleasing to children. Such biscuits, which are illustrated in Fig. 14, differ from cinnamon rolls only in the leavening agent used, cinnamon rolls being made with yeast and pinwheel biscuits with baking powder.

### Pinwheel Biscuits

(*Sufficient to Serve Six*)

2 c. flour

1 tsp. salt 4 tsp. baking powder

2 Tb. fat f 3/4 c. milk 2 Tb. butter 1/3 c. sugar 1 Tb. cinnamon 3/4 c. chopped raisins

To make the dough, combine the ingredients in the same way as for baking-powder biscuits. Roll it on a well-floured board until it is about 1/4 inch thick and twice as long as it is wide. Spread the surface with the 2 tablespoonfuls of butter. Mix the sugar and cinnamon and sprinkle them evenly over the buttered surface, and on top of this sprinkle the chopped raisins. Start with one of the long edges and roll the dough carefully toward the opposite long edge, as shown in Fig. 15. Then cut the roll into slices 1 inch thick. Place these slices in a shallow pan with the cut edges down and the sides touching. Bake in a hot oven for about 20 minutes.

BEATEN BISCUITS.—In Fig. 16 is illustrated a form of hot bread known as beaten biscuits. Such biscuits are used very extensively in the South; in fact, they are usually considered typical of the South. Formerly, all the lightness of beaten biscuits was produced by beating, but as the mixture is made today it may be run through a food chopper a few times before it is beaten. If this is done, the labor of beating is lessened considerably, beating for 15 to 20 minutes being sufficient. When the beating is finished, the texture of the dough should be fine and close and the surface should be smooth and flat.

### Beaten Biscuits

(*Sufficient to Serve Twelve*)

1 qt. pastry flour

1 tsp. salt

1/3 c. fat

1 c. milk or water

Sift the flour and salt and chop in the fat. Moisten with the milk or water and form into a mass. Toss this on a floured board, and beat it with a rolling pin for 30 minutes, folding the dough over every few seconds. Roll the dough 1/3 inch in thickness, form the biscuits by cutting them out with a small round cutter, and prick each one several times with a fork. Place the biscuits on baking sheets or in shallow pans, and bake them in a moderate oven for 20 to 30 minutes.

### Miscellaneous Hot-bread Recipes

SOFT GINGERBREAD.—As a hot bread for breakfast, soft gingerbread like that illustrated in Fig. 17 is very satisfactory, and with or without icing it may be served as cake with fruit for luncheon. Sweet milk and baking powder are generally used in gingerbread, but sour milk may be substituted for sweet milk and soda in the proper proportion may be used in place of baking powder. If not too much spice is used in a bread of this kind, it is better for children than rich cake, and, as a rule, they are very fond of it.

## Soft Gingerbread

(*Sufficient for One Medium-Sized Loaf*)

2 c. flour

2 tsp. baking powder

1/2 tsp. soda

1/4 c. sugar

1/2 tsp. salt

2 tsp. ginger

1 tsp. cinnamon

1 egg

1/2 c. milk

1/2 c. molasses

1/4 c. butter or other fat

Mix the flour, baking powder, soda, sugar, salt, and spices. Beat the egg, add the milk and molasses to it, and stir these into the first mixture. Melt the fat and stir it into the batter. Pour the batter into a well-greased loaf pan, and bake in a moderate oven for about 35 minutes. If preferred, the mixture may be poured into individual muffin pans and baked in a moderate oven for about 25 minutes.

BOSTON BROWN BREAD.—A hot bread that finds favor with most persons is Boston brown bread.

Such bread, instead of being baked in the oven, is steamed for 3-1/2 hours. It may be made plain, according to the accompanying recipe, or, to give it variety, raisins or currants may be added to it. Boston brown bread may be steamed in an ordinary coffee can, such as is shown in Fig. 18, in a large baking-powder can, or in a can that is made especially for this purpose. A regular steaming can for Boston brown bread is, of course, very convenient, but the other cans mentioned are very satisfactory. A point to remember in the making of brown bread is that the time for steaming should never be decreased. Oversteaming will do no harm, but understeaming is liable to leave an unbaked place through the centre of the loaf.

## Boston Brown Bread

(*Sufficient for One Medium-Sized Loaf*)

1 c. white flour 1 c. graham flour

1 c. corn meal 3/4 tsp. soda

2 tsp. baking powder

1 tsp. salt

3/4 c. molasses

1-3/4 c. sweet milk

Mix and sift the flour, corn meal, soda, baking powder, and salt. Add the molasses and milk and mix all thoroughly. Grease a can and a cover that fits the can tightly. Fill the can two-thirds full of the mixture and cover it. Place it in a steamer and steam for 3-1/2 hours. Dry in a moderate oven for a few minutes before serving.

NUT LOAF.—The use of nuts in a hot bread increases the food value and imparts a very delicious flavour. It is therefore very attractive to most persons, but it is not a cheap food on account of the usual high price of nuts. Thin slices of nut bread spread with butter make very fine sandwiches, which are especially delicious when served with tea.

## Nut Loaf

(*Sufficient for One Medium-Sized Loaf*)

2 c. flour

1/2 c. sugar

4 tsp. baking powder

1 tsp. salt

4 Tb. fat

1 egg

1 c. milk

1/2 c. English walnuts

Mix and sift the flour, sugar, baking powder, and salt, and then work in the fat. Add the egg, well beaten, and the milk, and then stir in the nut meats, which should be chopped. Turn into a well-greased loaf pan, and bake in a moderate oven for about 45 minutes.

## Utilising Left-over Hot Breads

As a general rule, not much consideration need be given to the utilising of left-over hot breads, for these are not often baked in large quantities and consequently are usually eaten at the meal for which they are intended. Still, if any should be left over, they should never be wasted, for there are various ways in which they may be used. The small varieties, such as muffins, biscuits, etc, may be freshened so that they will be almost as good as when first baked by putting them into a hot oven for a few minutes. If they are quite stale, they should be dipped quickly into hot water before being placed in the oven. The moisture on the surface is driven into the interior of the bread by the intense heat, with the result that the biscuits become moist and appear as fresh as they did formerly. If it is not desired to freshen them in this way, biscuits, muffins, and even pieces of corn bread that have become slightly stale may be made delicious by splitting them and then toasting them.

## Luncheon Menu

As in the preceding Sections, there is here submitted a menu that should be worked out and reported on at the same time that the answers to the Examination Questions are sent in. This menu is planned to serve six persons, but, as in the case of the other menus, it may be increased or decreased to meet requirements. The recipe for macaroni with cheese and tomatoes may be found in *Cereals,* and that for baking-powder biscuit, as well as that for popovers with apple sauce, in this Section. Recipes for the remainder of the items follow the menu.

## RECIPES

### Watercress-and-celery Salad

Arrange on each salad plate a bed of watercress, or, if it is impossible to obtain this, shred lettuce by cutting it in narrow strips across the leaf and use it instead of the watercress. Dice one or two stems of celery, depending on the size, and place the diced pieces on top of the watercress or the lettuce. Pour over each serving about 2 teaspoonfuls of French dressing made as follows:

1/2 tsp. salt 1/4 tsp. pepper 1/4 tsp. paprika 6 Tb. oil 2 Tb. vinegar

Mix the salt, pepper, and paprika, and beat the oil into them until it forms an emulsion. Add the vinegar gradually, a few drops at a time, and continue the beating. Four the dressing over the salad.

**Tea**

Measure 1 teaspoonful of tea for each cupful that is to be served. Scald the teapot, put the tea into it, and add the required number of cups of freshly boiling water. Allow it to steep until the desired strength is obtained. Serve at once, or pour from the leaves, serving cream and sugar with it if desired.

# 7

# Other Important Food

RICE OR CHOWL

Rice is consumed by most European families at breakfast, tiffin, and dinner. It is eaten at breakfast with fried meat, fish, omelet, country captain, or some other curried dish, and, being invariably followed by toast and eggs, jams, fruit, &c., one /coonkee/ , which contains about as much as an ordinary breakfast-cup, or say half a pound, will always be ample for four tolerably hearty consumers. There are two sizes of /coonkees/, large and small: reference is here made to the /small coonkee/, well filled. The quantity, however, of raw rice for a party of four should not exceed half a pound.

The rice at dinner is usually preceded by soup, fish, roast, and made dishes.

The best or generally approved qualities of rice for table use are known as the bhaktoolsee, the banafool, the bassmuttee, and cheeneesuckur. In purchasing these, or indeed any other approved quality, care must be taken to avoid /new rice/ and what is called /urruah/, which latter has been put through some process of boiling, or damped, and then dried. Both are considered unwholesome for general daily consumption, and few Indians will use them.

Good rice when rubbed in the palm of the hand, and cleared of dust, will appear of a bright and nearly transparent yellowish colour; whereas the /urruah/ will be found of a dull whitish hue, and the grain streaked and speckled with white powder, which crumbles on the application of a needle's point.

The price of rice, like other commodities, varies according to its plenty or scarcity in the market. After the cyclone of October,

1864, and again of November, 1867, the price of the bhaktoolsee and the banafool, which are fine, large, stout-grain rice, without being coarse, ruled at from eight to nine seers per rupee, and the bassmuttee and the cheeneesuckur at from seven to eight seers per rupee. The rice used by the poorest classes of the native population is of a very coarse description and incredibly cheap: within six weeks after the cyclone of November, 1867, it was readily procurable at twenty-five to thirty seers per rupee.

Rice is used in a variety of forms: it is boiled, made into kitcheeree, pellow, puddings, blanc mange, cakes, bread, &c.

The bhaktoolsee, the banafool, and other stout-grain rice are the best adapted for boiling. Boiled rice is called /bhath/.

The bassmuttee, cheeneesuckur, and all small and fine-grain rice are selected for kitcheeree, pellow, and puddings for children's food, and for invalids.

The /urruah/ is used in some houses in ignorance, but for the most part it is made into flour, and used for blanc mange, cakes, &c. The flour is abundantly procurable in the Calcutta markets, and is largely used by all native bakers in the making of bread.

Twenty-two to twenty-five seers of rice monthly, consuming it three times a day, entertainments included, will be ample for a party of four, allowing occasionally for a rice pudding.

It is necessary to wash rice thoroughly in several waters before using it, and a colander is very useful for draining away the water after washing the rice.

## Boiled Rice

Wash half a pound or a coonkeeful of rice, and put it to boil in a large quantity of water, over a brisk fire. Immediately the rice begins to boil, the water will bubble up to the surface of the pot and overflow, carrying away quantities of scum and impurities. The cover of the pot should now be kept partially open, and the rice stirred to prevent an entire overflow of the water. On the subsiding of the water or the bubbling, the fire should be reduced, until it is satisfactorily ascertained that the grains of rice, without being pappy, are quite soft, when the pot should be removed from the fire and a quart of cold water be added. All the liquid, which is "conjee," should then be drained, and the pot replaced over a gentle charcoal heat, to allow all moisture to evaporate, assisting the process by occasionally shaking the pot, or stirring its contents gently with a wooden spoon. Time to boil: half an hour.

The coonkee of rice when properly boiled will fill a good-sized curry or vegetable dish. The rice will be found quite soft, and yet every grain perfectly separate. Rice should never be cooked into a pap, excepting it is required for very young children; and leaving the grains hard or uncooked should be equally avoided.

A small pinch of pounded alum or /fitkerree/ is used by some cooks with advantage to improve the whiteness of boiled rice.

## Rice Conjee

The water in which rice is boiled should never be thrown away: it is nutritious and fattening for all cattle, horses included, and may be given daily to milch cows and goats with great advantage.

## Rice Kheer

This is occasionally served upon the breakfast-table as a treat, but few Europeans care for it. It is made as follows:—Thoroughly boil one coonkee or half a pound of the bassmuttee or the cheeneesuckur rice, then drain the water away, add two cups of pure cow's milk, and put over a slow fire.

As the rice begins to absorb the milk, two or three small sticks of cinnamon are put in, with one tablespoonful and a half to two tablespoonfuls of fine-quality white sugar. On the milk being entirely absorbed, the kheer is either turned out upon a dish and eaten hot, or put into a buttered mould, served up in shape, and eaten cold.

Kheer is sometimes cooked or boiled in milk only, but the foregoing recipe is supposed to be that more generally approved.

## Pish-Pash

Pick and wash in several waters a coonkee or half a pound of the bassmuttee or other fine-grain rice; add to it, cleaned and cut up, a chicken, some sliced ginger, sliced onions, a few bay-leaves, some peppercorns, a few hotspice, a dessertspoonful of salt, one chittack or two ounces of butter, and water sufficient to cover the whole. Simmer over a slow fire until the chicken becomes perfectly tender and the rice quite pappy. Serve up hot. This is considered a most excellent and nutritious meal for invalids.

## KITCHEEREES

These are occasionally substituted for boiled rice at breakfast, and are eaten with fried fish, omelets, croquets, jhal frezee, &c. They are prepared as follows:—

### Bhoonee Kitcheeree

Take rather more than three-quarters of a coonkee of bassmuttee or cheeneesuckur and half a coonkee of dal; or, if preferred, take the rice and dal in equal parts.

Take twelve large curry onions and cut them up lengthways into fine slices. Warm up two chittacks or four ounces of ghee (but before doing so be careful to warm the pot), and, while bubbling, throw in the sliced onions, removing them immediately they become of a bright brown colour. Set the fried onions aside, and throw in the dal and rice (having previously allowed all the water in which they were washed to drain through a colander). Fry until the dal and rice have absorbed all the ghee; then add a few slices of green ginger, some peppercorns, salt to taste (say one dessertspoonful), a few cloves, three or four cardamoms, half a dozen bay-leaves, and as many small sticks of cinnamon. Mix well together; add as much water only as will entirely cover over the whole of the rice and dal, put a good-fitting cover on, and set over a slow fire, reducing the same from time to time as the water is being absorbed. Care must be taken not to allow the kitcheeree to burn, which may be prevented by occasionally shaking the pot, or stirring its contents with a wooden spoon.

Serve up quite hot, strewing over it the fried onions, which serve both as a relish and garnish of the dish.

### Bhoonee Kitcheeree of The Mussoor or Red Dal

Is made according to recipe No. 5.

### Bhoonee Kitcheeree of The Moong or Small-grain Yellow Dal

Is made according to recipe No. 5.

### Bhoonee Kitcheeree of The Gram or Chunna Dal

The chunna or gram dal makes a very nice kitcheeree; but, as it is rather hard, it should be boiled or soaked in cold water for an hour or so before frying it with the raw rice.

### Bhoonee Kitcheeree of Green Peas

Kitcheeree made of green peas grown of English seeds is a rarity. Large peas should be picked out and shelled; they should not be fried with the rice, but added to it when nearly cooked. The instructions given in recipe No. 5 are to be observed in all other respects.

### Jurrud or Yellow-tinted Kitcheeree

Jurrud or yellow-tinted kitcheeree is nothing more than one of the above kitcheerees, to which is added, at the time of frying the rice and dal, either a small quantity of saffron or turmeric, according to the colour desired to be imparted. Such introduction in no way affects the flavour, nor does it render the appearance of the dish more attractive, but serves admirably as a variety for a large breakfast-table.

### Geela Kitcheeree

This is usually made of moong dal with less than one-fourth the quantity of ghee allowed for the bhoonee, or with no ghee at all, and little or no condiments are used, excepting a small quantity of finely-sliced green ginger, a few peppercorns, one or two bay-leaves, and salt to taste. It is supposed to be better adapted than bhoonee kitcheeree for children and invalids.

By /bhoonee/ is meant crisp, and /geela/ signifies soft.

## PELLOW OR POOLOO

Pellows are purely Hindoostanee dishes. There are several kinds of pellow, but some of them are so entirely of an Asiatic character and a taste that no European will ever be persuaded to partake of them. It is therefore considered useless to offer instructions how to prepare such as the /ukhnee pellow/, in which are introduced cream, milk, butter-milk, garlic, and lime-juice; or the /sweet pellow/, in which almonds and raisins are introduced, in addition to sugar, &c.

The following are the pellows in general use:—

### Chicken Pellow

Take a good-sized chicken; clean, truss, and boil it with one pound of beef in two cupfuls of clean water, seasoning it with onions, ginger, and salt. When sufficiently cooked, but yet quite firm, remove the chicken, and set it and the gravy aside. Cut up twelve onions lengthways into fine slices. Warm your pot; then melt in it two chittacks or four ounces of ghee, and, as it bubbles, throw in the sliced onions and fry to a light brown; remove and set aside. Then put in half a pound, or a coonkee, or the best bassmuttee or cheeneesuckur, having drained away all the water in which it was washed, and fry. On the rice absorbing the ghee, throw in a few cloves, four or five cardamoms, half a dozen small sticks of cinnamon, some peppercorns, a blade or two of mace, and one dessertspoonful of salt. Mix up the whole, and pour over it the gravy in which the chicken and beef were boiled, or as much of it only as will entirely cover the rice; close the pot immediately with a close-fitting cover, and set on a slow fire. As the gravy continues to decrease or to be absorbed, so keep reducing the fire, shaking up the pot occasionally, or stirring its contents, to prevent the pellow from burning. Brown the boiled chicken in a pan with ghee or butter, and serve up as follows:—

Place the chicken, either whole or cut up, on the centre of a dish, covering it with the pellow; strew over it the fried onions, garnishing it besides with two hard-boiled eggs, cut into halves, or in some device, and with half a dozen bits of finely-sliced and fried bacon, to suit the taste of those who like the latter.

## Beef, Mutton, or Kid Pellow

Take two pounds of beef, and cut up as for a curry, or take a small but good leg of mutton, or two legs of a kid, rejecting the loin.

Make a good, strong gravy with seasoning of sliced onions, ginger, and salt, with water, which when cooked down will be reduced to about sufficient only to cover the rice. Then proceed to make the pellow in all respects as directed in the foregoing recipe. The beef is not further used for the table, but treat the legs of the kid, or the mutton, the same as the chicken, and serve up with fried onions, hard-boiled eggs, and fried bacon, like the chicken pellow.

## Prawn Pellow

Instead of a chicken, provide yourself with eight or ten good-sized "bagda prawns," and a good hard cocoanut. After frying and setting aside the sliced onions, as directed above, the rice is to be fried, but, instead of using chicken or any other meat broth, cook it in the milk of the cocoanut (/vide/ recipe No. 54), observing in all particulars the instructions given for the chicken pellow, recipe No. 12, and serve up as follows:—Dish up the pellow, strew over it the fried onions, and garnish with the prawns finely boiled, and two hard-boiled eggs cut in halves or in some other device.

The cocoanut milk will impart a sweetish flavour to the pellow, but it is not disagreeable; and its sweetness may be subdued, if required, by reducing the strength of the cocoanut milk.

### Lobster or Fish Pellow

Take out the centre bones or one or two hilsa or beckty fishes, which are procurable fresh and good in the market, and eight or ten large long-legged lobsters with the roe or coral; thoroughly wash in several waters with salt, and boil with plenty of seasoning of onions, sliced ginger, peppercorns, a dozen bay-leaves, a tablespoonful of unroasted dhuniah or coriander seed, and salt, with water sufficient to give the required quantity of gravy. When ready, remove and shell the lobsters, reserving the roe or red coral in the heads, which bruise down with a little unroasted coriander seed, and mix with the fish gravy. Make the pellow in all other respects the same as prawn pellow, using the gravy of the fish instead of cocoanut or other gravy, and garnish with the lobsters, &c.

## CURRIES

A curry-stone and muller, or what the natives call /seal our lurriah/, are necessary for the preparation of condiments for daily use. The condiments should be carefully, and each kind separately, ground down to a nice paste with a little water.

Condiments prepared with water will not keep good any number of days; if required for a journey, therefore, or as presents for friends at home, good sweet oil and the best English vinegar should be substituted for the water. For the preparation of condiments for this purpose see recipe No. 65.

The first cost of a curry-stone and muller of large size will not exceed one rupee, but they will require re-cutting every three or four months, at a cost not exceeding one anna each re-setting.

The following is a list of curry condiments and hotspice in almost daily use:—

Curry onions, or /carree ka piaj/,price from 3 to 8 pice per seer.

| | |
|---|---|
| Turmeric, or /huldee/ | " 3 to 5 annas " |
| Garlic, or /lussoon/ | " 2 to 3 annas " |
| Green ginger, or /uddruck/ | " 2 to 4 annas " |
| Dry chilies, or /sooka mirritch/ | " 3 to 5 annas " |
| Coriander-seed, or /dhunnia/ | " 3 to 4 annas " |
| Cumin-seed, or /jeerah/ | " 5 to 6 annas " |
| Peppercorns, or /gool mirritch/ | " 5 to 6 annas " |
| Bay-leaves, or /tage paththa/ | " 2 to 3 annas " |
| Lemon-grass, or /uggheaghass/ | " 3 to 6 pice for a bundle of 16 to 20 blades of grass. |
| Poppy-seed, or /post ka danna/ | " 3 to 4 annas per seer. |
| Onion-seed, or /cullinga/ | " 5 to 8 annas " |

Stick cinnamon, or /dalcheenee/

Cardamoms, or /elachee/

Mixed; prices range from Rs. +/-3.14 to 4 per seer.

Cloves, or /loung/

Mixed; prices range from Rs. +/-3.14 to 4 per seer.

Nutmeg, or /jyephall/

Mixed; prices range from Rs. +/-3.14 to 4 per seer.

Mace, or /jowttree/

Mixed; prices range from Rs. +/-3.14 to 4 per seer.

However high prices may range, one rupee-worth of mixed condiments, including hotspice, will suffice for a month's consumption for a party of from four to six adults, allowing for three curries per day, cutlets and made dishes included.

## GRAVY CURRIES

The following directions for an every-day gravy chicken curry will apply equally to all ordinary meat gravy curries:—

### Chicken Curry

Take one chittack or two ounces of ghee, two breakfast-cupfuls of water, one teaspoonful and a half of salt, four teaspoonfuls of ground onions, one teaspoonful each of ground turmeric and chilies, half a teaspoonful of ground ginger, and a quarter of a teaspoonful of ground garlic.

To suit the taste of those who like it, half a teaspoonful of ground coriander-seed may be added, which should be roasted before being ground. Observe the following directions for cooking:—

Take the usual full-sized curry chicken, the price of which has latterly ranged from three to four annas, and divide it into sixteen or eighteen pieces. Warm the pot, melt in it the ghee, and immediately it begins to bubble throw in all the ground condiments, stirring until quite brown; then put in the cut-up chicken and the salt, and stir up to a good light-brown colour; then add the water, and allow the whole to simmer over a slow fire until the chicken is quite tender, and the liquid reduced to about half its original quantity. The operation of cooking or simmering will take from a half to three-quarters of an hour.

### Kid Curry

Take a hind-quarter or a fore-quarter of kid, which may be obtained at from three to four annas the quarter; cut it up into sixteen or eighteen pieces; take condiments in the proportion given in recipe No. 16, and cook it in every particular the same as the chicken curry, allowing it to simmer three-quarters of an hour.

### Veal Curry

A small shoulder of veal, the price of which ranges from three to four annas, may be selected; cut off from it sixteen or eighteen one-inch square pieces of the best part of the meat, and curry it in every particular the same as a chicken, only allowing it to simmer half to three-quarters of an hour.

### Mutton Curry

Obtain a small shoulder at from five to six annas; cut it up into sixteen or eighteen one-inch square pieces, rejecting all the bones; curry it the same as a chicken, allowing it to simmer for half an hour longer, or until the meat is tender.

N.B.—The bones of the veal and mutton, referred to in this and the foregoing recipe, may be turned to account for stock or gravy for some made dish.

## Beef Curry

Two pounds of well-selected meat will cost from three to four annas; cut it up into one-inch square pieces, rejecting all the scraggy parts; cook it in every respect according to the instructions given in recipe No. 16 for cooking a gravy chicken curry, only allowing it to simmer for a much longer time than any other curry, or until the beef becomes tender.

## Green Duck Curry

The price of a young tender duck may be quoted at from four to five annas. Cut it up exactly as you would a chicken, and curry it in the same manner, allowing it to simmer for an hour and a half. It is desirable to introduce half a teaspoonful each of coriander and cumin seeds in this curry.

## Young Pigeon Curry

Take four young pigeons; cut each into four pieces, making in all sixteen pieces. The price of young pigeons ranges from five to six annas the pair. The instructions given for the cooking of a gravy chicken curry apply equally to a pigeon curry.

## DOOPIAJAS

The literal translation of /doopiaja/ is "two onions," and the term probably is correctly applicable, as it will be noticed, in the recipes for preparing the /doopiaja curries/, that besides the full quantity of ground onions, it is necessary to put in about an equal quantity of fried onions, thereby /doubling/ the quantity of onions.

Doopiajas are more piquant curries; they are cooked with more ghee and less water. The following condiments, &c., are considered ample for a really good /docpiaja/ of chicken or of any meat:— One chittack and a half or three ounces of ghee, one breakfast-cupful of water, one teaspoonful and a half of salt, four teaspoonfuls of ground onions, one teaspoonful each of ground turmeric and chilies, half a teaspoonful of ground ginger, a quarter of a teaspoonful of ground garlic, twelve onions cut lengthways, each into six or eight slices, and half a teaspoonful of ground coriander-seed if it be liked.

## Chicken Doopiaja

Take a full-sized curry chicken and divide it into sixteen or eighteen pieces. Melt the ghee in a warm or heated pot, fry brown the sliced onions and set aside; then fry the ground condiments, stirring the whole; when brown, add the cut-up chicken with the salt, and fry to a rich brown. Chop the fried onions and put into the pot with one cup of water, and allow to simmer over a slow fire for about one hour, when the chicken will be perfectly tender, and the liquid reduced to a thick consistency, and to half its original quantity.

### Kid Doopiaja

Is made in all respects as a chicken doopiaja, the kid to be cut up in the usual manner. The hind quarter is preferable to the fore quarter.

### Veal Doopiaja

Take only the meat from a shoulder, cut it up into squares, and allow it to simmer for half an hour longer than the chicken doopiaja.

### Mutton Doopiaja

The flesh part of a shoulder is cut up into squares and doopiajed exactly as a chicken, allowing it to simmer over a slow fire for half an hour longer.

### Beef Doopiaja

Cut two pounds of beef into one-inch square pieces, and follow all the instructions given in recipe No. 23, only allowing it to simmer for a much longer time over a slow fire, until the beef is perfectly tender.

### Duck Doopiaja

Divide as you would a chicken, and cook the duck in the same manner, allowing it to simmer a little longer than the chicken doopiaja. Half a teaspoonful each of ground coriander and cumin seed should be mixed with the condiments.

### Doopiaja of Pigeons

Take four pigeons, cut each into four pieces, and proceed in every particular the same as for a chicken doopiaja.

### Cold Boiled Pork Doopiaja

Cut from the remains of cold boiled pork sixteen one-inch square pieces, and doopiaje it in the way directed for a chicken.

The time required to simmer will not exceed that allowed for the chicken doopiaja.

### Udder Doopiaja

Take two pounds of udder; before cutting it into squares, it should be parboiled, and then made into doopiaja, allowing it to simmer over a slow fire for about two hours.

### Udder and Beef Doopiaja

Take one pound each of udder and beef; parboil the udder, and then cut it up with the beef into one-inch square pieces, and doopiaje it, allowing it to simmer for about two hours.

It is necessary to impress on the amateur artist the importance of paying particular attention to the firing: a brisk fire will dry up the ghee and the water before the curry is half cooked, and necessitate the addition of more water, which will in every instance spoil the doopiaja, although the addition of a little water, if such be necessary when the curry is nearly cooked, will do it no harm. In every instance where ghee butter, &c., is to be melted, it is desirable first to warm the pot.

## FORCEMEAT BALL CURRIES, OR COFTA-KA-CARREE

Beef, mutton, chicken, fish, crabs, and prawns are usually taken for making these curries. The ingredients for two pounds of meat or fish are as follow:—Lard, ghee, or mustard oil, three to four ounces; water or stock, five to six ounces; ground onions, one tablespoonful or one ounce; ground chilies, a quarter of a tablespoonful, ground turmeric, a quarter of a tablespoonful; ground green ginger, half a teaspoonful; ground peppercorns, half a teaspoonful; ground garlic, a quarter of a teaspoonful; garden herbs, finely chopped, one dessertspoonful; salt, one dessertspoonful; finely-grated bread-crumbs, three tablespoonfuls; one egg.

N.B.—In the fish, crab, and prawn coftas the ginger must be omitted.

### Beef Forcemeat Ball Curry

Get rather more than two pounds of good fat beef; wash it thoroughly, and cut it into pieces, rejecting all veins and scraggy portions; put about two pounds of it into a mortar and pound it fine, removing all fibres, veins, &c., and if it be desired put up a broth of all the rejections. Mix with the pounded beef a teaspoonful of salt, pepper, and garden herbs, and two tablespoonfuls of bread-crumbs; add a little of the broth, or in its absence some milk; mix the whole well together; beat up the yolk and white of the egg, add it to the mixture, and make into balls about the size of large walnuts; roll them in bread-crumbs. After heating the pot, melt the lard or ghee, and fry brown the ground ingredients, sprinkling a tablespoonful of cold water over them; then add the coftas or balls with salt to taste, and fry or brown them; after which pour into the pot either a cup of broth or of water, and allow to simmer for about two hours.

N.B.—Some cooks add to the beef cofta curries ground hot spices, which are fried with the curry condiments, and are suited to most tastes.

## Chicken Forcemeat Ball Curry

Procure a good fat chicken and a quarter of a pound of beef suet; put the suet into a mortar with all the fleshy parts of the chicken, and pound to a pulp; make a stock of gravy of the bones; mix with the pounded meat all the several ingredients named in the foregoing recipe, with the addition of an egg well beaten up; make into balls, roll in bread-crumbs, and curry as directed above.

N.B.—The chicken cofta curry may also be made without any suet; the general practice is to get chickens rather larger than those usually selected for ordinary curries.

## Mutton Forcemeat Ball Curry

Take the best parts of a leg or shoulder of mutton; cut them up, wash, and pound well down; make a gravy of the bones and rejections; mix with the pounded mutton all the ingredients mentioned in the recipe for making beef balls, and cook exactly as the beef cofta curry.

## Ball Curry of Liver and Udder

Get one pound each of liver and udder; thoroughly wash and parboil them, then cut them into pieces, put into a mortar, and pound them to a pulp; mix with pepper, salt, herbs, bread-crumbs, and an egg; make into balls, and curry them in the same manner as any of the foregoing forcemeat ball curries.

## Prawn Cofta Curry

Get thirty to forty of the best prawns, and remove the heads and shells; wash the prawns well with salt and water, then pound them to a pulp; mix with it all the ingredients as directed for the beef cofta; make into balls, roll them in bread-crumbs, and set aside.

After washing the heads, remove the shells, and bruise the contents with a dessertspoonful of unroasted coriander-seed; take all the juice, and fry it with the ground condiments; then put in the balls, brown them, add salt to taste, a cup of water, and simmer until they are cooked.

N.B.—Good mustard oil is preferable to using lard or ghee, and the ginger must be omitted; but the addition of a few bay-leaves and blades of lemon-grass would be an improvement. It is not usual to dish up the lemon-grass.

## Lobster Cofta Curry

According to their size, take eight or ten lobsters; clean them thoroughly; remove the heads and shells; pull the flesh to pieces and pound to a pulp; add to it some of the red coral from the head, then mix into it the bread-crumbs, salt, pepper, herbs, and an egg well beaten up, and make into balls. The remains of the heads and the contents of the long legs bruise down with unroasted coriander-seed; omitting the ginger, and cook the balls in the same way as the prawn balls, with the addition of bay-leaves and a few blades of lemon-grass. Lemon-grass is not served up.

## Crab Cofta Curry

Select ten or twelve /gheewalla kakakahs/, or crabs full of the red coral, wash them thoroughly, then boil them; remove all the meat and coral out of the shells, pound to a pulp, and, after mixing all the ingredients and fixing them with an egg well beaten up, make into balls, and cook them in all respects according to the directions for lobster cofta curry. Time to simmer: say half an hour.

## Fish Cofta Curry

Cold boiled or fried fish is the best adapted for making coftas; it is not necessary to give other instructions than those already given at length in the foregoing recipes, excepting that mustard oil is the best adapted for fresh fish curries.

N.B.—The remains of hermetically-sealed fish, such as salmon and mackerel, removed from dinner, are well adapted for making cofta curries.

Under-done roast meats, such as beef, mutton, veal, and fowl, will make excellent cofta curries.

## COUNTRY CAPTAIN

The country captain is usually made of chicken, and occasionally of kid and veal. Cold meats and curries are also sometimes converted into this dish, the condiments for which are as follow:—Two chittacks or four ounces of ghee, half a teaspoonful of ground chilies, one teaspoonful of salt, a quarter of a teaspoonful of ground turmeric, and twenty onions, cut up lengthways into fine slices.

### Chicken Country Captain

Cut up in the usual way an ordinary curry chicken. Warm the ghee and fry the sliced onions, which when brown set aside; fry the ground turmeric and chilies, then throw in the chicken and salt, and continue to fry, stirring the whole, until the chicken is tender. Serve it up, strewing over it the fried onions.

### Kid Country Captain

Before cutting up the kid, a fore-quarter, let it be partially broiled or roasted, and then make it into country captain in accordance with the above directions; or, instead of partially roasting the kid, add half a cup of water to assist the meat to dissolve.

### Veal Country Captain

Partially broil or roast a shoulder of veal before cutting it up; or make the country captain as directed in recipe No. 42, by adding half a cup of water instead of partially broiling the meat.

### Jhal Frezee

Cut up into small squares, of less than an inch, either cold mutton, beef, or veal, rejecting the bones; add a large quantity of sliced onions, some chilies cut up, and a teaspoonful of salt. Warm a chittack, or two ounces of ghee, and throw it into the meat, onions, chilies, and salt, and allow to simmer, or fry, stirring the whole while, until the onions are quite tender.

## HINDOOSTANEE CURRIES

### Seik Kawab

Is usually eaten with chappatee or hand-bread, and only occasionally with rice, and contains the following condiments:— Two tablespoonfuls of mustard oil, four teaspoonfuls of ground onions, one teaspoonful of ground chilies, half a teaspoonful of ground ginger, a quarter of a teaspoonful of ground garlic, one teaspoonful of ground turmeric, one teaspoonful and a half of salt, a cup of thick tyre or dhye, half a teaspoonful of ground coriander-seed, the juice of one large lemon, and a little ghee.

Take two pounds of beef, mutton, or veal; remove the bones, and chop the meat slightly, without mincing or cutting through it; mix well together all the ground condiments, including the oil, tyre, and lemon-juice, in which steep the chopped meat, turning it over occasionally to absorb the mixture. After a while cut up the meat into squares of equal size, say two inches, and continue to keep them in the mixture for fully one hour; then pass the squares of meat either on a silver, plated, or other metal skewer, and roast or broil over a slow charcoal fire, basting the whole time with ghee, to allow the kawab to become of a rich brown colour, without burning or being singed in the basting. Remove from the skewer, and serve hot.

### Tick-keeah Kawab

Take two pounds of fat beef, wash it, cut it into small pieces, and pound it to a pulp, remove all fibres, &c., and then add to it one teaspoonful of ground onion, a quarter of a teaspoonful of ground turmeric, one-eighth of a teaspoonful of ground garlic, a quarter of a teaspoonful of ground chilies, half a teaspoonful of ground peppercorns, a quarter of a teaspoonful of ground ginger, half a teaspoonful of ground hot spices, and one tablespoonful of tyre or dhye.

Mix the whole well together, add salt to your taste, and the yolk and white of an egg well beaten up; form into balls of equal sizes; flatten them, pass them on iron or plated skewers about eighteen inches long, rub them well over with ghee, wrap them in plantain-leaf, and roast or broil them over a charcoal fire. Serve them up hot, removed from the skewers. These are usually eaten with chappatee.

## HUSSANEE CURRIES, OR CURRIES ON STICK

The ingredients and condiments necessary for the curries on stick are as follow:—One chittack and a half of ghee, one teaspoonful and a half of salt, four teaspoonfuls of ground onions, one teaspoonful of turmeric, half a teaspoonful of ginger, half a cupful of water, a quarter of a teaspoonful of ground garlic, one teaspoonful of chilies, half a cupful of tyre or dhye, some finely-sliced ginger, and as many small curry onions cut into half as may be required.

Six silver pins five inches long, or, in the absence of these, six bamboo pins, are required.

### Hussanee Beef Curry

Cut up two pounds of beef into small squares not exceeding one inch, and pass them on the silver or bamboo pins alternately with half an onion and a slice of ginger. Half a dozen sticks with be ample for four hearty consumers.

Warm the ghee and brown the ground condiments; then put in the sticks of meat, and brown, stirring the whole; after this add the tyre and a little water, and allow to simmer over a slow fire for nearly two hours, when the curry will be ready. Serve up on a curry-dish without removing the sticks.

### Hussanee Mutton Curry

Remove the meat from a shoulder of mutton, and cut it into small squares; the same instructions will apply to the preparation of mutton curry on stick as those given for beef curry on stick. Time to simmer: half an hour.

### Hussanee Veal Curry

Cut squares enough from a shoulder of veal, and observe the instructions given in the foregoing recipe. Time to simmer: one hour.

### Hussanee Curry of Udder and Liver

The udder and liver should be parboiled before being cut up for passing on the sticks; but in all other respects the instructions given for the beaf and mutton curries on stick will apply to the udder and liver curry on stick. Time to simmer: fully one hour and a half.

## KURMA OR QUOREMA CURRY

This, without exception, is one of the richest of Hindoostanee curries, but it is quite unsuited to European taste, if made, according to the original recipe, of which the following is a copy-

### Quorema Curry, Plain

Take two pounds of mutton, one pound of tyre or dhye, two chittacks of garlic, one dam of cardamoms, four chittacks of bruised almonds, four mashas of saffron, the juice of five lemons, one pound of ghee, four chittacks of sliced onions, one dam of cloves, one chittack of pepper, four chittacks of cream, and a quarter of a teaspoonful of ground garlic.

The following is the recipe of the quorema curry usually put on a gentleman's table:—Two chittacks and a half or five ounces of ghee, one cup or eight ounces of good thick tyre, one teaspoonful of ground chilies, four teaspoonfuls of ground onions, one teaspoonful of coriander-seed, six small sticks of ground cinnamon, two or three blades of lemon-grass, one teaspoonful and a half of salt, a half teaspoonful of ground ginger, a quarter of a teaspoonful of ground garlic, eight or ten peppercorns, four or five ground cloves, five or six ground cardamoms, two or three bay-leaves, a quarter of a cup of water, the juice of one lemon, and twelve large onions cut lengthways into fine slices.

Take two pounds of good fat mutton, and cut it up into pieces nearly one inch and a half square. Warm the ghee, fry in it the sliced onions, and set aside; then fry all the ground condiments, including the ground hot spices.

When quite brown, throw in the mutton and salt, and allow the whole to brown, after which add the tyre, the hot spices with peppercorns and bay-leaves, the lemon-grass, the water, and the fried onions finely chopped; close the pot, and allow it to simmer over a gentle coal fire for about an hour and a half or two hours, by which time the kurma will be quite ready. The blades of lemon-grass are never dished up.

### Kid Quorema

Cut up a fore-quarter or a hind-quarter of a kid into eight or ten pieces, and cook it exactly as directed in the foregoing recipe. This is rather preferred to mutton quorema.

## Fowl Quorema

Take a young full-ground tender fowl; cut it up as for an ordinary curry, cook it with all the condiments in the proportions given, and observe all the directions laid down.

N.B.—Most Europeans give the preference to the fowl quorema.

## MALAY CURRIES

The condiments and other ingredients necessary are as follow:—One chittack or two ounces of ghee, one teaspoonful and a half of salt, four teaspoonfuls of ground onions, one teaspoonful of ground turmeric, one teaspoonful of ground chilies, half a teaspoonful of ground ginger, a quarter of a teaspoonful of ground garlic, the milk of a large cocoanut, say two cups, two blades of lemon-grass, three or four cloves, ground, three or four cardamoms, and as many small sticks of cinnamon, ground. The coriander and cumin seeds must on no account be put into malay curries, or the delicate flavour of the cocoanut will be destroyed.

It will be necessary to provide what the natives call a /narial-ka-khoornee/, which, interpreted, means "cocoanut scraper." It is a small circular flat piece of iron, about the size and thickness of a Spanish dollar, the edges being notched. It is of rude

construction, and fixed on a conveniently shaped wooden frame, also of rude construction. The best of the kind may be procured for two annas.

## Cocoanut Milk

The nut is scraped or rasped with the aid of the "khoornee" into very fine particles; it is then put into a deep vessel, and boiling water poured over it until the whole of the scraped cocoanut is covered. After allowing it to steep for ten or fifteen minutes, it is carefully strained through a clean napkin into another vessel or cup, the pulp is returned into the original vessel, and more boiling water is poured over it. This operation of steeping in boiling water and straining is continued until you have obtained the required quantity of the extracted milk of the cocoanut. The pulp is thrown away. If the cocoanut be a small one, or its nut not hard and deep, it will be necessary to provide a second cocoanut. Good cocoanuts are sold at an anna to an anna and a half a piece.

## Chicken Malay Gravy Curry with White Pumpkin or Cucumber

Take the usual full-sized curry chicken, and divide it as before directed; get either six cucumbers or a quarter of a white pumpkin; remove the green skin and the part containing the seeds, then cut it up into sixteen pieces of about two inches square, and steep in water.

Fry in the ghee all the ground condiments, including the ground hot spices; when brown, add the cut-up chicken and salt; fry to a fine bright light brown; then put in the pumpkin, having previously allowed all the water to drain away through a colander; pour in the two cups of cocoanut milk, the lemon-grass, and hot spices, and allow the whole to simmer over a slow fire for about half an hour, when the curry will be ready: the blade of lemon-grass is not dished up.

## Prawn Malay Gravy Curry with White Pumpkin or Cucumber

Select the bagda prawns (/bagda chingree/), whenever they are procurable, in preference to any other description. The shell and head are of a dark colour in comparison with what are called /jeel ka chingree/, the shell and head of which are very perceptibly several shades lighter than the /bagdas/.

It is impossible to quote any price as a guide, the fluctuation being almost incredible. Fine large prawns, not lobsters—prawns which, without their heads, would be about the size of the ordinary dried Normandy pippins sent out to this country for tarts—may be obtained one day at two annas for twenty, and the next day they will not be procurable at less than eight annas for the same number. This remark applies generally to fish of every description brought for sale into the Calcutta market.

With one other remark of importance, we shall proceed to the instructions necessary for the preparation of prawn malay gravy curry.

The prawns should be parboiled after removing the heads, to rid them, as the natives call it, of /besine/, which means all disagreeable character of fishy smell and taste.

As a rule, the heads of prawns should always be rejected, which, in the process of frying, absorb largely the ghee, and in the cooking dispel a liquid from their spongy formation.

In all other respects, the prawn malay gravy curry is cooked like the chicken malay gravy curry, omitting the ginger; but an additional blade or two of the lemon-grass would not be amiss, which, on the curry being dished, are thrown away.

### Chicken Malay Gravy Curry with Pulwal

Take a fat chicken, clean it, remove all the flesh and pound it to a pulp, and prepare it in every respect as directed in recipe No. 34 for a cofta curry, omitting the suet. Take a dozen large-sized pulwals, scrape or pare away the outer skin, split them down one side, extract all the seeds, &c., and throw the pulwals into cold water; wash and drain away all the water, then stuff them with the prepared forcemeat, tie them with fine sewing cotton, and cook them in the milk of the cocoanut.

### Prawn Malay Gravy Curry with Pulwal

Take bagda prawns; shell and clean them, pound to a pulp, and prepare as directed in recipe No. 37 for prawn cofta curry. Take a dozen pulwals, peel them finely, cut them open lengthways, clear them of all seeds, &c., wash and dry them, then stuff them with the prepared prawn mince; tie the pulwals with sewing cotton, and cook in cocoanut milk as directed in recipe No. 56.

### Chicken Malay Doopiaja

The condiments and ingredients are as follow:—One chittack and a half or three ounces of ghee, one teaspoonful and a half of salt, four teaspoonfuls of ground onions, one teaspoonful of ground turmeric, one teaspoonful of ground chilies, half a teaspoonful of ground ginger, a quarter of a teaspoonful of ground garlic, one cup of strong cocoanut milk, and one dozen onions cut lengthways into fine slices. Cut up the chicken in the usual manner, warm the ghee, fry and set aside the sliced onions, then fry brown the ground condiments, after which add the chicken and salt. When fried brown, pour in the cocoanut milk and the fried onions finely chopped, and allow to simmer over a slow fire; the Malay doopiaja will be ready in an hour.

### Prawn Malay Doopiaja

Take sixteen or twenty large bagda prawns, throw away the heads, parboil the prawns, and then doopiaje in all respects as for a chicken Malay doopiaja, omitting the ginger.

## PORTUGUESE CURRY (VINDALOO OR BINDALOO)

This well-known Portuguese curry can only be made properly of beef, pork, or duck. The following is a recipe of the vindaloo in general use-

Six ounces or three chittacks of ghee or lard, one tablespoonful of bruised garlic, one tablespoonful of ground garlic, one tablespoonful of ground ginger, two teaspoonfuls of ground chilies, one teaspoonful of roasted and ground coriander-seed, half a teaspoonful of roasted and ground cumin-seed, two or three bay-leaves, a few peppercorns, four or five cloves, roasted and ground, four or five cardamoms, roasted and ground, six small sticks of cinnamon, roasted and ground, with half a cup of good vinegar, to two pounds of pork or beef or a duck.

N.B.—The best vindaloo is that prepared with mustard oil.

### Beef Vindaloo

Cut up two pounds of fat beef into large squares, and steep them in the vinegar, together with half a teaspoonful of salt and all the ground condiments, from eighteen to twenty-four hours. Then warm the ghee or lard and throw in the meat, together with the condiments and vinegar in which it had been steeped, adding a few peppercorns and bay-leaves, and allow to simmer gently over a slow fire for two hours, or until the meat is perfectly tender, and serve up hot.

### Pork Vindaloo

Cut up two pounds of fat pork into large squares, and curry according to the directions given in the foregoing recipe, omitting the cloves, cardamoms, and cinnamon.

### Duck Vindaloo

Take a young, full-grown, but tender duck; cut it up as for a curry, and put it through the same course of pickling from eighteen to twenty-four hours before being cooked.

### Pickled Vindaloo (adapted as a Present to Friends at a Distance)

If the following instructions be carried out carefully, the vindaloo will keep good for months, and, if required, may be sent as an acceptable present to friends at home.

In order to keep it good sufficiently long to be sent home round the Cape, select the fattest parts of pork; satisfy yourself that the meat is fresh and sound, and that it has not been washed with water in the butcher's shop. Cut the meat into two-inch squares, wash thoroughly in vinegar (no water), rub over with the following condiments, and then steep them in really good English vinegar for twenty-four hours:—Garlic bruised, not ground down, dry ginger powdered, turmeric powdered, peppercorns roasted and powdered, coriander-seeds roasted and powdered, cumin-seeds roasted and powdered, and dry salt.

Melt a large quantity of the best mustard oil in an earthen pot, and, according to the quantity of meat, take additional condiments mentioned above, but in the proportion given in recipe No. 61; grind in vinegar, and fry in the oil; then put in the meat, and all the vinegar, &c., in which it had been stepped, together with some more salt, a little more vinegar, a few bay-leaves and peppercorns, and allow to simmer until the meat is quite tender. Remove from the fire and allow it to get quite cold; then put it into dry stone jars, with patent screw tops, well filled with plenty of the oil in which the vindaloo was cooked. Take care that all the meat is well covered over with oil, which latter ought to be at least from two to three inches above the meat in the jar. Screw down the lid, and cover it over with a good sound bladder to render it perfectly air-tight.

When required for use, take out only as much as will suffice, and simply warm it in a little of its own gravy.

### Curry Paste

Is likewise adapted for sending as a present to friends at home. It is made in the following manner:—Eight ounces of dhunnia, or coriander-seed, roasted; one ounce of jeerah, or cumin-seed, roasted; two ounces of huldee, or dry turmeric; two ounces of lal mirritch, dry chilies; two ounces of kala mirritch, black pepper, roasted; two ounces of rai, or mustard-seed; one ounce of soat, or dry ginger; one ounce of lussan, or garlic; four ounces of nimmuck, salt; four ounces of cheenee, or sugar; four ounces of chunna or gram dal without husk, and roasted. The above ingredients, in the proportions given, to be carefully pounded and ground down with the best English white wine vinegar to the consistency of a thick jelly; then warm some good sweet oil, and while bubbling fry in it the mixture until it is reduced to a paste; let it cool, and then bottle it.

N.B.—Great care must be taken not to use any water in the preparation, and mustard oil is better adapted than sweet oil for frying the mixture in.

## MADRAS MULLIGATAWNY CURRY

As this dish is usually served up and partaken of in the place of ordinary soup, reference will be made to it hereafter under the head of "Soups."

Before proceeding to remark on fish, vegetable, and peas curries, a few useful hints and suggestions may be offered on meat curries generally.

In many families the remains of cold meat, if not required for other purposes, are made into curry: cold roast or boiled mutton is admirably adapted for the purpose; and in ninety-nine cases out of a hundred consumers cannot tell the difference. If there be any difference or advantage, it is decidedly in favour of the cold meat: the roasting joints are always of a superior quality to meats sold under the designation of "curry meats."

The remains of cold roast beef make the best cofta curries, croquets, &c., and if the beef be under-done no fresh beef will make a better doopiaja.

Vegetables are sometimes put into gravy meat curries, never into doopiajas; but, as a rule, the introduction of vegetable into any meat curry is objectionable, from the fact that all vegetables in the process of boiling or cooking throw out a liquid, some more and some less: the potato throws out the least, but of a disagreeable character. It is true potatoes may be boiled before being put into a curry, but the piquancy and peculiarity of flavour looked for in a curry is so palpably destroyed that the innovation may be discovered with closed eyes. The introduction of vegetable into gravy fish curries, however, is no innovation, as the condiments used for the one answer for the other; both are cooked in oil, and the ginger omitted.

### Gravy Fish Curries

The condiments are as follows:—Mustard oil, one chittack or two ounces; water, two cups; four teaspoonfuls of ground onions, one teaspoonful of ground turmeric, one teaspoonful of ground chilies, and a quarter of a teaspoonful of garlic. It will be noticed that mustard oil is used instead of ghee, and no ginger.

Too much care cannot be observed in thoroughly cleaning, rubbing, and washing the fish in salt and water before cooking it for the table. Fish, if properly washed, when served up will never be offensive, unless it be bad when purchased.

## Hilsa Fish Gravy Curry

The head and tail are thrown away, and the fish cut into slices of rather more than half an inch thick; these should be washed in several waters with salt, to rid them of all "besine," before they are curried.

The acid of tamarind is considered an improvement, or "amchoor," which is sliced green mangoe dried in salt.

## Beckty Fish Gravy Curry

Is sliced and washed in salt like the hilsa before being cooked. It is not usual to put any acid in the beckty fish curry.

## Prawn Doopiaja

Take one chittack and a half of mustard oil, four teaspoonfuls of ground onions, one teaspoonful of ground turmeric, one teaspoonful of ground chilies, a quarter of a teaspoonful of garlic, twelve curry onions cut lengthways, each into six or eight slices, one cupful of water, and twelve large prawns.

Clean and thoroughly wash the prawns, rejecting the heads, or taking only their substance pounded and squeezed out with unroasted coriander-seed, and after parboiling the prawns make the doopiaja in all respects according to the ordinary mode.

## Sliced Hilsa Fish Fried in Curry Condiments

Take two teaspoonfuls of ground onions, one teaspoonful of ground chilies, two teaspoonfuls of salt, half a teaspoonful of ground turmeric, a quarter of a teaspoonful of ground garlic, and one chittack of mustard oil.

After slicing a hilsa in the manner directed for a curry, and having thoroughly cleaned and washed it with salt, rub into the slices all the ground condiments and the remaining salt, and allow them to remain for at least an hour. Warm the oil, and fry the slices of fish of a very light and bright brown. Serve up hot.

### Sliced Beckty Fish Fried in Curry Condiments

Slice, wash, and fry exactly as directed above. Fish served up in this manner is well suited to some European tastes, and makes an agreeable change to the ordinary mode of frying fish for breakfast.

### Egg Curry

Take six or eight eggs, boil hard, shell, cut into halves, and set them aside; take ghee, ground condiments, and sliced fried onions, in all respects the same as for a chicken doopiaja, and observe precisely the same method of cooking, keeping in mind the fact that, the eggs being already cooked or boiled, a smaller quantity of water and a shorter time to simmer will suffice.

### Egg Curry with Green Peas

This is a favourite curry with some families in winter, when the English green peas are procurable. The method of preparing it is exactly the same as recipe No. 71, allowing the curry to simmer until the peas are quite tender.

### Egg Curry, with Chunna Ka Dal

Parboil and dal, say half a cupful; curry the dal first; when about nearly cooked, throw in the hard-boiled eggs, and finish the simmering immediately the dal is soft or tender.

## CHAHKEES

Chahkee is a term applied to vegetable curries, some of which are deservedly popular, and one in particular, which many families have daily during the season the vegetables are procurable, and yet never tire of, viz.-

### Seam, Potato and Peas Chahkee

Take twenty seams, four new potatoes, and a quarter of a seer of green peas; divide each seam into three pieces, and throw into a bowl of water; divide each potato into four pieces, and throw into water; shell the peas, wash all thoroughly, put into a colander to drain, and cook with the following condiments:—One chittack and a half of mustard oil, four teaspoonfuls of ground onions, one teaspoonful of ground chilies, half a teaspoonful of ground turmeric, a quarter of a teaspoonful of ground garlic, one teaspoonful and a half of salt, and one cupful of water. Warm the oil, let it bubble well, and fry the ground condiments; when these are quite brown put in the vegetables and salt; let the whole fry, stirring it well; then add the water, and allow it to simmer over a slow fire until the vegetables are quite tender.

N.B.—A cauliflower may be added if required for a change.

### Pulwal, Potatoes and Torrie

Clean as much of the above three kinds of vegetables as will overfill a vegetable-dish, and make the chahkee in all respects as the foregoing.

### Red Pumpkin and Tamarind

A quarter of a red pumpkin and the pulp of two or three tamarinds will be enough. Dissolve the pulp of the tamarind in the water, and put it into the curry after the pumpkin has been fried.

### White Pumpkin and Tamarind

Chahkee it in the same way as the red pumpkin.

### White Pumpkin, Plain, Cut Small

It is not necessary to give any further instructions than those already given.

### Tomato with Tamarind

Take twenty tomatoes and the pulp of two or three tamarinds, and chahkee as directed for red pumpkin.

### Tomato, Plain

Chahkee twenty tomatoes according to instructions given for other chahkees.

N.B.—There is a fresh green herb called soa mattee, which is sometimes put into fish, vegetable, and other curries. Some Europeans like the flavour, and have it daily when procurable. Inquiry and trial are recommended.

## SAUG CURRIES

Half an anna's worth of any saug will suffice for a party of four, for curries made of greens, such as spinach, &c. The following condiments, &c., are used:—One chittack and a half of mustard oil, four teaspoonfuls of ground onions, one teaspoonful of ground chilies, half a teaspoonful of ground turmeric, a quarter of a teaspoonful of ground garlic, one teaspoonful and a half of ground salt, and one cupful of water.

### Red Saug and Omra

The omra should be peeled, and half fried if large. Great care must be taken to thoroughly clean and wash the greens. Put them into a colander and allow all the water to drain away. Then warm the oil, and fry the ground condiments; then the saug and omra, and when crisp add the water and cook over a slow fire until the greens and omra are tender.

### Red Saug, Omra and Shrimps

Observe in all respects the same process as that required in cooking without the shrimps, omitting the ginger.

### Red Saug and Prawns

The prawns should be parboiled, and then follow all the instructions in recipe No. 82.

### Green Saug with Prawns

Proceed in every particular as with the last.

### Danta Curry with Shrimps

The danta is a fine delicate long green pod which the horseradish-tree yields, and contains small peas; these pods are cut into lengths of three or four inches and cooked with shrimps. Beyond this explanation it is not necessary to enlarge upon the instructions already given.

### Khuttah Carree, or Acid Vegetable Curry

Take small quantities of all kinds of vegetables in season, but the best curry is that made of potatoes, kutchoo or artichoke, sweet potatoes or suckercund, carrots, red and white pumpkins, and tomatoes.

The vegetables should be cut into large pieces, and boiled in water with the following condiments:—Four teaspoonfuls of ground onions, one teaspoonful each of ground turmeric and chilies, a quarter of a teaspoonful of ground garlic, and one teaspoonful of roasted and ground coriander-seed.

Prepare two large cups of tamarind water, slightly sweetened with jaggry, strain through a sieve, and add the strained water to the boiled vegetables with a few fresh chilies. Then melt in a separate pot one chittack or two ounces of mustard oil. While the oil is bubbling, fry in it a teaspoonful of the collinga, or onion-seeds, and when sufficiently fried pour it over the boiled vegetables including the tamarind water. Close up the pot, and allow it to simmer for fifteen to twenty minutes, when it will be ready. It is eaten cold.

## BHAHJEES

By /bhahjee/ is meant fried. The two most generally approved vegetable bhahjees are those made of bringals and pulwals. The following are the condiments, &c., used:—Mustard oil according to the quantity of vegetable to be fried, a little ground turmeric and chilies, and some salt.

### Bringal Bhahjee

Take young full-sized bringals; wash them thoroughly, and slice them about an eighth of an inch thick; dry them, steep them for half an hour in the ground condiments and salt, fry in oil, and serve up hot.

### Pulwal Bhahjee

Take a dozen or more pulwals—a most excellent and wholesome native vegetable,—scrape or pare away very finely the upper green coating, divide them lengthways into two pieces, clear away all the seeds, &c., wash, drain away all the water, and steep them in ground turmeric, chilies, and salt for half an hour or longer; then fry them quite crisp in melted mustard oil. They are much liked by some Europeans.

N.B.—The vegetable called ram's horns or lady's fingers, known by the natives as /dharus/, makes an excellent bhahjee; so does the /kerrella/, a small green and intensely bitter native vegetable, which comes into the market in March and April; it is not, however, well suited to the European taste.

## DAL OR PEAS CURRIES

Half an anna's worth of any dal will suffice for a party of four. The condiments are as follow:—Three-quarters of a chittack of ghee, four teaspoonfuls of ground onions, one teaspoonful of ground chilies, half a teaspoonful of ground turmeric, half a teaspoonful of ground ginger, a quarter of a teaspoonful of ground garlic, one teaspoonful and a half of salt, and half a dozen onions cut into six or eight slices each.

### Moong Dal

Take half a pound of the raw dal, or say half a cupful; clean, pick, and roast it; mix it up with all the ground condiments and salt, put into a pot, pour water over the whole, some two inches above the dal, and boil it well, until the dal has quite dissolved. Be careful not to disturb it while in the process of boiling, but allow it to cake as it were /en masse/. When thoroughly boiled, churn the dal by twirling it in a wooden instrument called a / ghootnee/; then warm the ghee in a separate pot, fry the onions, chop them, and throw into the churned dal, after which pour the dal into the pot of melted ghee, and keep stirring until the dal and ghee have well mixed; then put the cover on, and allow to simmer over a slow fire for about a quarter of an hour.

N.B.—The standard price of the best roasted moong dal is two annas and a half per seer.

### Mussoor or Red Dal

The process in all respects for preparing and cooking the red dal is the same as for the moong dal, excepting that, instead of fried sliced onions, a large clove of garlic is cut up small, fried, and takes the place of the onions. The price of the best quality mussoor dal, free of husk, is two annas per seer.

### Mussoor Dal with Amchoor or with Tamarind

Put the amchoor, or, if preferred, tamarind, into the pot with the dal; allow it to dissolve, and when the dal is going through the process of bring churned remove the hard stones of the amchoor or seeds of the tamarind.

### Mussoor Dal Chur Churree

Instead of only half a dozen onions, take a dozen, and cut them into fine slices lengthways. Warm the three-quarters of a chittack of ghee, fry and set aside the sliced onions, then fry all the ground condiments; next put in and fry the dal, having previously washed it well, soaked it in water for about a quarter of an hour, and drained it through a colander. When thoroughly fried and browned, add only a little water, barely sufficient to cover the fried dal, and allow to simmer from ten to fifteen minutes, or until the dal has dissolved. Serve up, strewing over it the fried onions. If chunna ka dal be used, soak it for an hour. Other dals are occasionally served up, but very rarely at European tables. The price of the best clean chunna ka dal rarely exceeds two annas per seer.

### Dal Foolaree

Is much liked by Europeans, but is rarely served up well, owing to the trouble and time required in making it properly.

## BURTAS OR MASHES

Burtas are mashes of potatoes and other vegetables, cold meats, dry fish, &c.; they are palatable, and much liked by most Europeans as accompaniments to curry and rice. The ingredients to almost every burta are the fine large white Patna onions, fresh green chilies, and the juice of fresh lemons.

### Potato Burta

Take a moderate or middling sized white Panta onion; remove the outer coats, and slice very fine; then slice or cut up two hot green chilies, and squeeze over the onion and chilies the juice of a fresh lime: allow to soak. Take eight or ten well-boiled potatoes, half a teaspoonful of salt, and a teaspoonful of good mustard oil; bruise the potatoes down with a large silver or plated fork, adding, when they are half bruised, the onions and chilies, with as much only of the lime-juice as may be agreeable: mix all well together with a light hand, so that the potatoes may not cake, and yet be well and thoroughly mashed and mixed.

### Brinjal Burta

Prepare the sliced onions, chilies, and lime-juice in the manner directed for potato burta. Take two fine young brinjals of large size; carefully and thoroughly roast them in a quick ash fire; remove the ashes and burnt parts of the skin, if any; then open the brinjals, and with a clean spoon remove the contents to as near the skin as possible, to which add a good teaspoonful of salt and teaspoonful of mustard oil; work these with a spoon to a perfect pulp, throwing away the lumps or shreds if any; then mix with it all the onions, chilies, and lime-juice. If not to your taste, add more salt or lime-juice, according to fancy.

### Dry Fish Burta

Prepare onions, chilies, and lime-juice as before. Take a part of the Arabian dried beckty and well broil it; remove all the bones, and pound the fish to nearly a powder; mix it thoroughly with a teaspoonful of mustard oil, and add the onions, chilies, and lime-juice.

### Red Herring Burta

Take onions, chilies, and lime-juice. Place the herring, with its original paper packing, on a gridiron, or on a frying-pan, and warm it well; then clear it of all skin, very carefully pick out /all/ the bones, bruise the herring, and mix it thoroughly with the sliced onions, chilies, and lime-juice.

N.B.—This is an inimitable burta.

### Cold Corned-Beef Burta

Steep sliced onions and chilies in lime-juice; have the red well-corned part of a cold round of beef nicely pounded; add to it the onions, chilies, and as much of the lime-juice as may be desirable.

### Cold Tongue Burta

The remains of a well-corned cold tongue make an excellent burta, as per recipe for cold beef burta.

### Cold Ham Burta

Is made in the same way as the beef and tongue burtas.

### Green Mango Burta

The condiments for this burta are a quarter of a teaspoonful of ground chilies, half a teaspoonful of ground fresh mint-leaves, half a teaspoonful of ground ginger, half a teaspoonful of salt, and a teaspoonful of sugar.

Take two ordinary large-size green mangoes; peel, divide, and throw them into clean water, remove the stones, then bruise them to a perfect pulp with the aid of the curry-stone and muller. Care must be taken that the stone is perfectly clean, and will not impart the flavour of garlic or turmeric to the burta. Mix the sugar well with the pulp; if the mango be very acid, add a little more sugar; then mix it with the salt and ground condiments; more salt or sugar may be added if required.

### Tomato Burta

Bake in an oven a dozen good-sized tomatoes until the skin cracks; break them down, and mix with them a little ground chilies, ginger, salt, and half a teaspoonful of good mustard oil. A small squeeze of lemon-juice may be added if desired.

## SOUPS

A digester is best adapted for boiling soups in, as no steam can escape, and consequently less water is required than in a common pot.

To extract the substance or essence of meat, long and slow boiling over a charcoal fire is absolutely necessary. In the first instance, however, it is desirable to boil up the meat with pepper and salt on a quick, brisk fire, and take away all the black scum which rises to the surface; then pour a little cold water into the pot to raise up the white scum, which also remove, and reduce the fire, taking care that in the process of slow-boiling the pot is never off the boil.

### Shin of Beef Soup

Take a shin of beef, cut it up small, wash it thoroughly, and boil with pepper and salt in sufficient water to well cover the meat. Let it boil over a brisk fire, taking away the black scum; add a little cold water, and skim off the white scum; then reduce the fire, and allow the soup to simmer until it somewhat thickens; strain the soup, cut away all the fat, season with soup herbs, and add more pepper and salt if necessary. Give it a good boil up, and then clear it with the white of an egg well beaten up, after which add a tablespoonful of Lea & Perrin's Worcestershire sauce, and half a wineglassful of sherry.

### Shin of Beef Soup, with Forcemeat and Egg Balls

Prepare a shin of beef soup in all respects according to the above directions; clear with an egg well beaten up, add to it sauce, sherry, forcemeat, and egg balls.

### Vermicelli Soup

Prepare a shin of beef soup as directed above, omitting the sauce and sherry. Parboil some vermicelli, and after clearing the soup with the white of an egg, add to it the parboiled vermicelli, and give it all a good boil up before serving.

## Macaroni Soup

Prepare a shin of beef soup as directed above, omitting the sauce and wine; boil some macaroni until perfectly tender; clear the soup with the white of an egg, then add the boiled macaroni, and warm up before serving.

## Mulligatawny Soup

Prepare a shin of beef soup as above, omitting the sauce, wine, and white of egg; set the soup aside. Take a full-sized curry chicken; cut it up into sixteen or eighteen pieces, and wash them thoroughly. Warm a pot and melt it into two chittacks or four ounces of ghee; fry in it some finely-sliced onions, and set aside. Then fry in the melted ghee the following condiments, &c.:—Four teaspoonfuls of ground onions, one teaspoonful of ground turmeric, one teaspoonful of ground chilies, half a teaspoonful of ground ginger, a quarter of a teaspoonful of ground garlic, half a teaspoonful of roasted and ground coriander-seed, and a quarter of a teaspoonful of roasted and ground cumin-seed.

Sprinkle a little water over these while frying; then add the cut-up chicken with two teaspoonfuls of salt. When nearly brown, add one chittack or two ounces of roasted and ground poppy-seeds; pour in the beef soup, add the fried onion and half a dozen of the /kurreah fool/ leaves, close the pot, and allow the whole to simmer over a slow fire until the chicken be perfectly tender. Serve up hot, with limes cut in slices on a separate plate.

## Mulligatawny Soup-Another Way

Prepare a shin of beef as directed above. Cut up a chicken; wash it and set it aside. Heat a pot and melt in it two chittacks or four ounces of ghee. After frying in it and setting aside some finely-sliced onions, fry the condiments in the proportions given in the foregoing recipe; then add the cut-up chicken with two teaspoonfuls of salt; brown it nicely; have ready two chittacks or four ounces of roasted and ground chunna ka or gram dal, which mix thoroughly in a cup of strong cocoanut milk, and pour over the chicken just as it has become brown; stir it well, and add the fried onions and the soup, with half a dozen of the /kurreah fool/ leaves; close the pot, and allow the whole to simmer for three-quarters of an hour. Serve up hot, with limes, either whole or cut in slices, on a separate plate.

## Delicious Curry Soup

Prepare a strong beef soup; slice some onions, and cut up a chicken; take curry condiments as directed above, omitting the coriander and cumin seed; melt two chittacks or four ounces of ghee; fry and set aside the sliced onions, then fry the condiments, add the cut-up chicken, and fry that also. In a part of the beef soup boil a spoonful of tamarind, so as to separate the stocks and stones; strain and stir it into the fried chicken: After a while add the remainder of the beef soup, with half a dozen /kurreah fool/ leaves, and the friend onions; close up the pot, and continue to simmer the whole until the chicken is quite tender. Serve up hot.

## Bright Onion Soup

Take a shoulder of veal; cut it up small, breaking all the bones; wash it thoroughly, put it into a pan with pepper, salt, and water, boil it well, and remove all the scum as it rises; reduce the fire, and let it simmer until the meat is perfectly dissolved; strain it, cut away all the fat, add soup herbs, and more pepper and salt if required; give it a boil up, and clear it with the white of an egg well beaten up; slice very fine some pure silvery white Patna onions, and steep them in boiling water, changing the water three or four times, every ten minutes; drain away all the water and add the onions to the soup; boil, and serve up hot.

## Bridal Soup, or Soup Elegant

Take two large shoulders of veal; cut them up small, bones and all, and, after washing thoroughly, boil over a brisk fire, with /white/ pepper and the best white salt. Be careful that the scum that rises is well skimmed; reduce the fire, and allow it to simmer until the meat falls off the bones; strain the soup, let it cool, and then thoroughly free it of all fat; return it into a clean digester, add more salt and white pepper if necessary, and some /white/ stocks of celery; boil it, and clear it with the whites of two eggs well beaten up; strain through flannel and set aside.

Take the best and most transparent parts of a calf's head and the tongue, and boil perfectly tender without reducing them to shreds, being careful to remove all the scum that rises to the surface; lay the boiled tongue and meat out on a clean dish; slice the tongue fine, and cut out all manner of devices, such as, diamonds, squares, circles, hearts, stars, &c.; do the same with the best and cleanest parts of meat selected from the head; take care that no particles of scum or other impurities be adhering to them; where any does adhere, rinse it off in a little of the cleared soup; then place them carefully into the tureen in which it is purposed to serve up the soup. If fancy macaroni be procurable, a tablespoonful may be boiled tender, free of all particles of dust or powder, and added to the cut-up meat and tongue, over which pour the boiling-hot soup; add to it a wineglassful of the palest sherry, and serve up hot.

N.B.—The calf's tongue and meat of the head may be boiled with the veal, but they should be removed when tender, and not allowed to dissolve with the longer simmering of the veal.

This is an elegant soup, beautifully transparent, and of the colour of light champagne.

### Soup Royal

Take a shin of beef, the best parts of meat cut off from a calf's head, and the tongue; cut the beef into small particles, but leave the tongue and the meat from the calf's head whole; add pepper and salt, and boil well, clearing the scum as it rises; remove the tongue and the meat of the calf's head when sufficiently tender, but continue to boil the shin of beef until it is well dissolved; then strain it, and cut away all the fat; put it up again with plenty of soup herbs, and more salt and pepper if necessary; boil it well up; squeeze into the soup the juice of half a lemon, and skim it well; strain it once more, and set it aside.

Cut the tongue into slices of an eighth of an inch thick, trim them into the shape of large diamonds, and set aside. Cut up the meat of the calf's head into one-inch squares and strips of an inch and a half long and half an inch wide; add to these a few ready-fried circular flat brain cakes, make also a few egg balls and forcemeat balls, and, after cooking, add them to the rest of the meat, tongue, &c., and set aside.

Take four red carrots, one pound of green peas, half a pound of boiled potatoes, one large turnip, one large Patna onion, a quarter of a pound of roasted and ground split peas or gram dal, some soup herbs, pepper, and salt, the pulp of one orange, and the peels of half an orange and half a lemon. Put these into a stewpan with water sufficient to cover the whole; boil them thoroughly, skimming all the while; when perfectly dissolved, turn them out into a colander and allow all the water to drain away; then turn the contents of the colander into a sieve, and pass the vegetables, &c., through it, rejecting all such as will not pass. Add the whole, or a part of the strained vegetables to the soup, which should not be thicker in consistency than a good thick potato soup.

Next stew one dozen good French prunes in a claret-glassful of port wine, which also strain through a sieve, rejecting stones, &c., and add the strained portion to the soup; then boil the whole, strain it once more, add to it all the forcemeat and egg balls, the brain cakes, tongue, &c., and serve up, adding to it more salt, wine, or sauce, if needed.

N.B.—This soup properly made is without its equal.

## FISH

### Fish Mooloo

Fry the fish and let it cool. Scrape a cocoanut, put a teacupful of hot water into it, rub it well, strain it and put aside; then put two spoonfuls more of water; strain this also; cut up three or four green chilies, and as many onions as you like, with half a garlic. Fry them with a little ghee, and whilst frying put the last straining of the cocoanut water in with the ingredients till it is dry; then add the first water of the nut, and pour the whole over the fish, with some vinegar, ginger, whole pepper, and salt to your taste.

### Fish Mooloo-Another Way

Fry in a little ghee three or four chilies cut up, half a clove of garlic, and some sliced onions. When half fried, add two tablespoonfuls of cocoanut milk, and continue to fry until dry; then stir into it a teacupful of cocoanut milk, a little vinegar, some sliced ginger, peppercorns, and salt to taste, and while hot pour it over a cold fried or boiled fish.

### Fish Mooloo and Another Way

Cut up a fish into small, two-inch squares, and fry in ghee, with egg, bread-crumbs, and turmeric, of a nice brown colour. Boil in cocoanut milk some sliced green ginger and sliced green chilies; then add the fish, with salt to taste, and let it stew until the sauce has thickened. Serve up hot.

## Prawn Cutlet

Shell and wash the prawns; remove the heads, but leave the tails; slit them down in the centre, and gently beat them flat with a rolling-pin; sprinkle them with pepper and salt, and some finely-minced soup herbs; rub them over with yolk of eggs, and dredge with flour; fry over a very moderate fire to a rich light brown colour. Garnish the dish with fried green parsley, or serve up with tomato sauce gravy as per recipe No. 300.

## Crabs in Shell

Clean and boil the crabs in salt; remove them out of the shells; pick and clean them well, and reserve the coral for dressing.

Chop and mince fine the crabs; add some onion and ginger juice, a little lime-juice, pepper, and salt, and a little mushroom catsup. Melt some butter, and fry the mixture in it until the butter be absorbed; then add a little stock, and remove from the fire immediately the stock begins to dry. Butter the shells, and fill with the mixture. The meat of six crabs will refill five shells. Take some finely-sifted bread-crumbs; grind down the coral, and put it over the mixture on the shells, with the bread-crumbs, and bits of butter; bake for a few minutes.

## Tamarind Fish

Make a thick pickle of ripe tamarinds, good English vinegar, and a little salt; pass through a sieve, rejecting all stones and fibres. Select really good fresh hilsa fish, full size, with roes. Remove all the scales and fins, cut away the heads and tails, remove the roes, clean out the fish inside, and then slice up, an inch thick. Wipe away all blood, &c., with a clean dry towel. Care must be taken to use no water in the cleaning of the fish or in the preparation of the pickle. The board on which the fish is cut up, and also the knife, must be very clean. After all the blood, &c., has been thoroughly cleaned and wiped away, lay out the slices of fish and roe on a clean dish, sprinkle thickly with salt, and place over them a wire dish-cover to keep away the flies. Four or five hours afterwards put a layer of the pickle into a wide-mouthed bottle or jar, and a thick coating of pickle over each slice of fish and the roes, after washing away the salt with a little vinegar; lay them in order in the jar, until the last of the fish is put in; then be careful to pot in a very thick layer of the pickle. Cork the jar securely, and tie it down with a good bladder to keep it air-tight, and in three weeks it will be fit for use. It is desirable to fill each jar well up to the mouth, to effect which the jars or bottles to be selected should be of the required size.

N.B.—If the fish be really fresh, all the ingredients of good quality, and no water used in the operation of cleaning and pickling, the jars well filled, and mouths secured with sound bladder, the fish will keep good for months, and will be fit to send home.

## Smoked Fish

The mango fish, beckty, or hilsa should be cut down the back, spread open, and well washed and salted. Have a bright charcoal fire, and sprinkle over it some bran, with brown sugar; cover the fire with an open-work bamboo basket, having over it a coarse duster; arrange the fish over the duster, and allow them to smoke. When one side has browned, turn and brown the other side. As the smoke decreases, add more bran, and fan up the fire. A duster thrown over the fish while smoking will facilitate the operation.

## Dried Prawns

Strip the prawns of their shells; keep them for a day in salt mixed with turmeric; then string and put them out in the sun daily for fifteen or twenty days.

## Prawn Powder

Take a seer of dry prawns; wash them well, dry over the fire until crisp, pound fine, with some red pepper and nutmeg, pass through a sieve, and bottle for use. A teaspoonful spread over bread and butter is considered a relish.

# JOINTS, MADE DISHES, ETC.

## Corned Round of Beef

Select a good round of beef four days previously to it being required for the table, together with two seers of cooking salt, eight fresh juicy limes, one anna-worth of saltpetre, and a tablespoonful of suckur, a description of moist brown sugar. Pound fine the saltpetre; put the rind of four limes, pared fine, into a marble mortar, with a tablespoonful of brandy or other spirit; bruise and pound it well, adding to it the suckur or brown sugar, and gradually half the powdered saltpetre; mix all well together.

Take one seer of the salt, and mix into it the contents of the marble mortar; divide the mixture into four equal parts, and rub briskly one-fourth part of it into the round; puncture the beef lightly during the operation with a clean bright steel sailmaker's needle, to allow the mixture to penetrate more freely. An hour or two after take another fourth of the mixture; squeeze into it the juice of the four limes from which the rind had been removed, and repeat the operation of rubbing it into the round, puncturing it lightly with the needle; turn the beef over from side to side continually, so that one side do not soak or steep more in the brine than another; repeat the operation of rubbing it well several times during the day. Next morning place it on a dry dish, and rub into it another fourth part of the prepared salt; let it stand for an hour or so, then pour over it the old brine; repeat the rubbing two or three times during the day, turning the beef continually. On the third day rub half of the remaining saltpetre into the beef dry, and allow it to stand for an hour or two; then add the rest of the saltpetre and the juice of the four limes to the remaining fourth part of the mixture, in which keep turning and rubbing the beef during the day as before; in the evening pour over it the stale brine, cover it thickly with the one seer of remaining salt, and place a heavy weight upon it, until required to be boiled the next day.

## Beef a la Mode

Corn a round of beef in every particular as directed above, and twenty-four hours previously to its being cooked lard it as follows with the undermentioned ingredients:—Four pounds of lard or fat bacon, half a tablespoonful of cinnamon powdered, half a seer or one pound of finely-powdered pepper, one tablespoonful of cloves powdered, and four tablespoonfuls of chutnee strained through muslin. Mix the ground pepper, ground hot spices, and strained chutnee with a claret-glassful of mixed sauces, such as Harvey, walnut, Worcestershire, tap, tomato, &c. Cut up into long narrow slips the lard or bacon to correspond in thickness with the larding-pin, and lay the slips into the mixture of spices, sauces, &c., for an hour or two before larding the beef, which should be larded through and through, and as closely as possible.

Cook it the next day, either in plain water, with half a pint of vinegar, and with bay-leaves and peppercorns, as is usual, or in a preparation of claret or champagne with vinegar, bay-leaves,

&c. This is not necessary, but it tends to the improvement of the flavour at some considerable cost.

### Le Fricandeau de Veau

Take a large leg of veal; remove the knuckle-bone; corn and lard it in all respects like a beef à la mode, reducing the ingredients in proportion to the difference in size and weight between a round of beef and the leg of veal. Boil, baste, and glaze it well in the liquor in which it is boiled. Serve up with all sorts of boiled and glazed vegetables.

### Hunter's Beef, or Spiced Beef

Corn a round of beef, as per recipe No. 123, with the addition of large quantities of finely-ground pepper and hot spices. Some of the pepper and spice should be well rubbed in with the saltpetre, and the beef should be punctured well the whole time with a needle to insure the saltpetre and spices penetrating.

After the dry saltpetre and spice have been well rubbed in, prepare a mixture of salt, saltpetre, suckur, lemon-rind, pepper, and spice, and rub in one-fourth of the mixture, continuing to puncture the beef. Add subsequently to the brine the juice of lemon, and observe closely all the instructions given in recipe No. 123. On the seventh day remove the beef from the brine; rub it well with two tablespoonfuls of finely-powdered spices and pepper; inclose it thoroughly in skins of fat, and then in a strong coarse pie-crust, and bake it in a good oven. A baker's oven is the best.

### Collared Brisket

Bone a brisket of beef; rub into it saltpetre, suckur or brown sugar, and one seer of salt, with some lime-juice; keep it in the brine for thirty-six hours, rubbing it continually. Then remove it from the brine, and clear away all the salt. Roll the beef tightly into a collar, secure it well, inclose it in a stout duster, and boil it.

### Spiced Collared Brisket

The process is the same as the above, but if the beef be required to keep for any lengthened time the quantity of salt ought to be doubled, the beef kept in the brine for seventy-two hours, and hot spices, pepper, chutnee, and sauces added. The beef after being rolled should be packed in the skin of fat, then in a coarse pastry, instead of in plantain-leaf, and baked in a baker's oven.

### Pigeons with Petit Pois

Kill and feather, with plunging into hot water, four young, full-grown pigeons, taking care not to break their skins; singe them, to destroy any remaining feathers; then wash them in three or four cold waters, cut them in halves, dredge them well with salt and finely-sifted pepper, and allow to remain for an hour. Then boil up two tablespoonfuls of ghee or lard, and fry the birds to a rich brown, turning them over. When sufficiently browned, put in a cupful of beef stock, and allow to simmer until the birds are quite tender; pour over them a tin of petit pois with their gravy, and serve up hot.

### Ducks with Green Olives

Choose young, full-grown, tender ducks; feather and singe them as directed in the foregoing recipe, after which wash them in three or four cold waters; stuff the ducks according to recipe No. 325, and bake in a deep dish in a moderate oven until brown; then add a good beef stock with sliced onions, and bake until the stock is reduced; remove the ducks, and put into the pan the contents of a bottle of olives stoned, and allow to bake for ten or fifteen minutes to soften the olives; place the ducks on a clean dish, arrange the olives round the ducks, and pour the gravy over. Serve up hot.

### Kidney Stew

Steep in lukewarm water for a few minutes a dozen mutton kidneys, and remove the white skin or coat which will become perceptible; cut into halves or quarter them, wash in three or four waters, and allow them to remain as long as possible in pepper, salt, and the juice of onions, ginger, and garlic; boil up three dessertspoonfuls of ghee or lard in a deep frying-pan, throw in the kidneys with the juice, put in half a clove of garlic, and cover over the whole with eight large Patna onions sliced each into eight slices, and separated so as to cover over the whole surface of the pan; pour over it as much hot stock as will keep all the onions under, and simmer over a slow fire until the onions disappear, when serve up quite hot.

### French Mutton Chops

Take half a dozen chops cut from a breast of mutton, throwing away the intermediate bones—that is to say, allow the meat of two chops to remain on one bone. Wash, dry, and steep the chops for an hour or two before dinner in the juice of onions, ginger, and garlic—say four teaspoonfuls of the first to three of the second and two of the last. Mix on a large board pepper, salt, and flour, with which dredge the chops thoroughly, and fry quickly in boiling ghee or lard, taking care in turning over and removing the chops not to use a fork or anything likely to occasion any wound to the chops, which should be held by the bones with a pair of pincers. Serve up hot immediately they have become of a good rich brown colour.

### Mutton Stew

Cut up a breast of mutton in the usual way for a stew; wash and dry the meat. Take of the juice of onions one tablespoonful, of ginger half a tablespoonful, and of garlic a quarter of a tablespoonful; mix with the meat, add pepper and salt, and allow to stand for any time from one to four hours.

Fry in a large stewpan two tablespoonfuls of ghee or lard, and when on the boil fry to a nice brown all the meat only; afterwards pour in the liquor in which the meat has been steeped, and allow to simmer for fifteen or twenty minutes; thicken some stock with a teaspoonful of flour, and add it to the stew; allow to simmer until the meat is perfectly tender.

If vegetables be required (the addition of which, however, is not considered any improvement), the original gravy, before adding the stock, must be removed and set aside.

Let the vegetables, consisting of, say, potatoes, carrots, turnips, and cut-up and sliced cabbage, after being cleaned, remain for an hour or two in cold water; lay them over the meat, and pour in hot stock sufficient to cover the whole of the meat and vegetables, and allow to simmer over a brisk coal fire until quite tender; then pour into the pot the original gravy which had been removed, and serve up hot. Or, instead of the vegetables named above, take only twenty-five or thirty tomatoes, in which case the stock should be lessened, as the tomatoes produce a large amount of liquid, and do not require as much boiling as the harder vegetables.

## Mutton Brains and Love Apples

Take six brains, sixteen to twenty large tomatoes, two chittacks or four ounces of butter, and eight biscuits. Wash the brains well; clean, boil, and halve, or cut each into three pieces; thoroughly butter the dish which will be put on the table; dredge it well with finely-powdered biscuit; lay in the brains; cut the tomatoes, and lay them in the dish between the brains, the cut ends upwards; add a small cupful of good stock, and, after sprinkling a sufficient quantity of pepper and salt as a seasoning, dredge thickly over with the ground biscuit-powder, and bake of a rich brown. Serve up hot.

## Kid Roasted Whole

Bespeak from a butcher a whole kid, with its head on.

Prepare a stuffing as per recipe No. 323 or 325, and after cleaning the kid, stuff into it the stuffing; break the joints of the legs, and fold and truss them like a pig; then put it up to roast, basting it the whole time with beef suet melted down, to which add hot water and salt. Serve up in a sitting posture like a pig, and with a lime in the mouth.

## Potato Pie

Boil and mash down some potatoes, with pepper, salt, milk, and butter; line a pie-dish a quarter of an inch thick with the mash; arrange in it a nicely-browned mutton, beef, or chicken stew, cover it over with a thick coat of the mashed potatoes, and bake for a quarter of an hour.

## Minced Veal Potato Pie

Make a good rich veal mince, mixed with a little ham, and some sippets of bread-crumb cut into small squares, diamonds, &c., and fried in butter; line the pie-dish with mashed potatoes as above directed; fill into it the veal mince, with plenty of gravy; arrange the sippets, cover over with a thick crust of the mashed potatoes, and bake for a quarter of an hour.

## Beef Steak and Pigeon Pie

This should consist of a slice of good steak, two pounds of beef, one chittack or two ounces of ghee, a teaspoonful of salt, two fresh limes, four young pigeons, twelve oysters, twelve curry onions cut lengthways into fine slices, a teaspoonful of ground pepper, some sweet herbs, and a dessertspoonful of flour.

Cut up the steak into pieces three inches long, and two inches or two and a half wide, by half an inch thick. Cut and divide each pigeon into four pieces; put up two pounds of beef with sufficient water to make a good strong gravy, throwing in all the scraggy parts and other rejections of the steak and pigeons. Warm the ghee, and fry in it the sliced onions; throw in, well dredged with the flour, the steaks and pigeons, and after frying a while add the pepper, salt, soup herbs, and some of the rind of the limes, and about half the beef gravy. Set the whole on a slow fire, and simmer until the meat is tender; allow to cool; then add the oysters and the remaining gravy, with the juice of two limes; put into a dish lined with pastry, cover over the whole with a pastry crust, and bake.

### Veal Pie

Cut a leg of veal into small pieces, or a breast into chops, and parboil in water enough to fill the pie-dish. When about half stewed take the veal out; season the gravy with pepper, salt, a little mace, and a little bacon; dredge in a little flour; line the sides of the dish with a pie-crust; arrange the meat, pour in the gravy, cover it with a pie-crust, and bake it for an hour.

### Macaroni Pie

Take half a pound of macaroni; boil and throw away the first water; then boil it again in some milk, and remove when it is quite tender. Prepare a strong gravy or soup with two pounds of beef, well seasoned with ground white pepper, salt, and soup herbs.

Bruise into fine powder two ounces of some good English cheese; take a dessertspoonful of very dry mustard, half a teaspoonful of very finely powdered white pepper, about two teaspoonfuls of salt, and two chittacks or four ounces of butter. Pound very fine a couple of crisp biscuits.

Pour over the boiled macaroni sufficient beef gravy or stock to entirely cover it; then put in all the pepper, salt, and mustard, but only half the ground cheese. Set it to simmer over a slow fire until the gravy begins to dry, and the macaroni acquires some consistency. Then with three ounces of butter (free of water) butter well the baking-dish; pour into it the macaroni; mix the remaining ground cheese with the powered biscuit, and strew it over the pie; cut into small pieces the remaining ounce of butter, and throw that also over the pie; then put the dish into an oven, and bake to a fine light but rich brown colour. Ten to fifteen minutes' baking will be sufficient.

### Alderman's Mock Turtle Pie

Make an extra rich hash of a calf's head, cutting the pieces from the cheeks two and a half to three inches long, and one and three-quarters to two inches wide. Slice the tongue, and cut into large-sized shapes. Prepare brain cakes, and plenty of forcemeat and egg balls.

Make an extra strong stock with eight calves' feet; season it highly with soup herbs, salt, and plenty of ground black pepper; simmer until the meat begins to drop away from the bones; strain through a coarse sieve, in order to get a very thick stock, passing as much of the dissolved meat through as possible.

Line a deep pie-dish with a thick and rich pie-pastry, and arrange in it the hash, egg and meat balls, and brain cakes, with some twenty or thirty green leaves of spinach, cut up to about the size and shape of the meat. Pour over the whole as much stock as will fill the dish, cover over with pastry, and bake.

### Sauce for Alderman's Mock Turtle Pie

Mix with some of the stock the contents of a canister of oysters well bruised, the pulp of sixteen or twenty prunes, a blade of mace, some nutmeg and cloves, a wineglassful of port wine, and a tablespoonful of Worcestershire sauce; allow to simmer for ten minutes, and add it to the ready-baked pie before it is put on the table.

### Friar Tuck's Mock Venison Pastry Pie

Take the chop ends of two large fat breasts of mutton; remove the bones, and after the meat has been washed, cleaned, and dried, lard well with narrow slips of lean bacon and corned tongue; then cut it up into twelve well-shaped chops nicely trimmed; steep them in the juice of onions, ginger, and garlic in the proportion of one tablespoonful of the former to a dessertspoonful of the latter, and half a teaspoonful of the last.

Make a strong broth or stock of the other side of the mutton, and all the rejections of bones, &c.; season it well with pepper, salt, and soup herbs; remove the scum and cut away all the fat; then strain through a sieve, rejecting all the fat, but passing through some of the lean; allow it to simmer until it thickens, then add to it two blades of mace, half a dozen allspice, and as many small sticks of cinnamon.

Line a deep metal pie-dish with the pastry pie-crust as per recipe No. 200, reserving sufficient for the upper crust. Prepare a sausage roll, say six inches long, and two inches and a quarter thick, of minced veal and udder, using the ordinary pie-crust pastry to inclose it in; then slice it into twelve equal slices of the thickness of half an inch.

Remove the twelve chops out of the onion, garlic, and ginger juice; dredge them well with finely-sifted flour mixed with pepper and salt; heat in a large deep frying-pan four tablespoonfuls of lard; fry the chops of a light brown colour, and remove them carefully; then dredge with flour and slightly brown the twelve slices of sausage, six of which lay at the bottom of the pie-dish; lay over them six of the mutton chops; over the mutton chops place another layer of the sliced sausage roll, and over that the remaining six chops; pour in as much of the stock or gravy as will fill the pie-dish, cover it over with a layer of the pastry as per recipe No. 200, and bake carefully.

### Sauce for Friar Tuck's Mock Venison Pastry Pie

Put some of the stock or gravy into the pan in which the chops and sliced sausages had been browned; add two tablespoonfuls of bruised and powdered oysters, and simmer from ten to fifteen minutes. Serve hot, on the pie being cut, adding at the last moment a wineglassful of port wine and one tablespoonful of lime-juice. Make a hole in the centre of the pie through the crusts, and pour in the sauce with the help of a lipped sauce-boat.

### Leg of Mutton Dumpling

Prepare a good pie-crust with one seer and a quarter of soojee, half a seer of flour, and half a seer of suet, as per recipe No. 199. Clean and trim the leg, cutting away the end of the knuckle-bone, and any other projections likely to injure the dumpling. Sprinkle it well with ground pepper and some salt, and confine it securely in the pastry, closing all joinings with the aid of a little water. Place the dumpling into a strong napkin, previously buttered and dredged with flour; tie it securely, and allow it to boil from three to four hours. Care must be taken that during the whole process of boiling the dumpling remains suspended in the water, and not resting on the bottom of the pan. On removing it from the boiler, plunge it immediately into a large tureen of cold water for two or three minutes. This will strengthen the pastry and prevent its bursting or breaking while it is being served up.

## Sausage Rolls

Take equal portions of cold roast veal and ham, or cold fowl and tongue; chop them together very small; season with a teaspoonful of powdered sweet herbs, and a spoonful of mixed salt and cayenne pepper; mix well together. Put three tablespoonfuls of the meat well rolled together into enough pastry (pie-crust recipe No. 199) to cover it. When you have used up the whole of your materials, bake them for half an hour in a brisk oven. These rolls are excellent eating, either hot or cold, and are especially adapted for travelling, gipsy, boating, or pic-nic parties.

## Dumpode Goose (Indian Way)

Take a good fat tender goose; feather, clean, and bone it carefully without destroying the skin; when every bone has been removed, pour into the goose a mixture composed of a dessertspoonful each of mustard, sweet oil, and mixed sauce.

Take all the bones and the giblet, the liver excepted, and make a good gravy seasoned with pepper, salt, soup herbs, and bay-leaves. Mince very fine three pounds of beef, a quarter of a pound of beef suet, a quarter of a pound of fat bacon, and the liver of the goose. Take of chopped garden herbs a tablespoonful, powdered black pepper a dessertspoonful, mixed hot spices finely powdered a dessertspoonful, finely-grated bread-crumbs two tablespoonfuls, salt a dessertspoonful, and essence of anchovies, if liked, one teaspoonful. Mix the above well together, and stuff the goose.

Melt two chittacks and a half or five ounces of ghee; put in the goose, and pour over it the soup made of the bones and giblet, and allow it to stew until quite tender; then glace the goose, as also some boiled turnips, carrots, onions, and potatoes, and serve up hot, surrounded with the vegetables and some English pickles.

## Dumpode Duck (Eastern Way)

Take a good fat duck; feather, clean, and bone it without hurting the skin; pour into it a mixture made up of a teaspoonful each of mustard, sweet oil, and mixed sauce.

Make a gravy of the bones and giblet, seasoning it with pepper, salt, soup herbs, and a few bay-leaves.

Mince together with the liver of the duck two pounds and a half of good beef, half a pound of beef suet, a dessertspoonful of chopped garden herbs, a tablespoonful of grated bread-crumbs, half a teaspoonful of mixed hot spices pounded, a teaspoonful each of black pepper and salt, and, for those who like it, half a teaspoonful of essence of anchovies. Mix these well together, and stuff the duck. Melt one chittack and a half or three ounces of ghee; put in the duck; pour over it the giblet gravy, and allow it to cook until tender; then glace the duck, as also some ready-boiled mixed vegetables, and serve up, surrounding the duck with the vegetables and some hot West-Indian pickle.

## Fowl a la Cardinal, or Dumpode Capon or Fowl

Feather the bird, clean it, and remove every bone very carefully without injuring the skin.

Make a good strong broth or gravy of the bone and giblet, reserving the liver.

Pour into the bird a mixture of sweet oil, mustard, and sauces in the proportion of one teaspoonful of each.

Mince the liver together with one pound and a half of good beef, one pound and a half of beef suet, a dessertspoonful each of finely-chopped garden herbs and finely-grated bread-crumbs, a teaspoonful each of powdered mixed hot spices, finely-powdered black pepper, and salt, if liked, and half a dozen oysters. Mix all well together, and stuff the bird; melt two chittacks or four ounces of ghee, add to it the giblet gravy, cook and glace the bird in it, as also some vegetables, and serve up hot, adding a little cayenne pepper to the gravy to make it piquant.

## Brisket of Beef Trambland

Heat or melt in a saucepan two chittacks of butter free of water; fry to a light brown a tablespoonful of finely-sliced onions, then add a tablespoonful and a half of flour, which must be put in very gradually, stirring the whole time; add half a teaspoonful of ground pepper, and one teaspoonful of salt. When these have been well mixed, pour in gradually a large cupful of pure milk, and lastly two wineglassfuls of vinegar. Keep stirring to prevent the sauce lumping. Mince fine half a dozen pickled gherkins or French beans, and mince up also the yolks and whites of six hard-boiled eggs. Boil a fresh brisket of beef, and dish up quite hot. Pour over it the sauce, over which sprinkle the minced pickle, and then cover the whole with the minced eggs.

## Mutton Trambland

Is prepared, in all respects, as the above. The joint best adapted to "trambland" is a fore-quarter, or only the shoulder, or the breast if required for a small party of two or three.

## Bubble and Squeak

Put into a pot cold meat cut into thin slices two inches square, with ready-boiled peas, cauliflower, cabbage, potatoes,

turnips, and carrots cut up, with pepper, salt, and sliced ginger, and with as much good stock as will cover the meat and vegetables; allow the whole to simmer until the meat and vegetables have absorbed half the stock, when it will be ready. Serve it up bubbling and squeaking.

## To Stew a Fillet of Veal

Bone, lard, and stuff a fillet of veal; half roast, and then stew it with two quarts of white stock, a teaspoonful of lemon pickle, and one of mushroom catsup. Before serving, strain the gravy; thicken it with butter rolled in flour; add a little cayenne, salt, and some pickled mushrooms; heat it, and pour it over the veal. Have ready two or three dozens of forcemeat balls to put round it and upon the top. Garnish with cut lemon.

## Veal Cutlets

Cut a neck of veal into cutlets, or take them off a leg. Season two well-beaten eggs with pounded mace, nutmeg, salt, pepper, and finely-chopped sweet marjoram, lemon, thyme, and parsley; dip the cutlets into it; sift over them grated bread, and fry them in clarified butter. Serve with a white sauce, forcemeat balls, and small mushrooms. Garnish with fried parsley.

### Kidney Toasts

Pound, in a marble mortar, the kidney and the surrounding fat; season with pepper, salt, grated lemon-peel, and nutmeg; mix with it the yolk of an egg well beaten; lay it upon thin toasted bread cut into square bits; put a little butter into a dish, lay in it the kidney toasts, and brown them in an oven. Serve them very hot.

### Rolled Mutton

Bone a shoulder of mutton carefully, so as not to injure the skin; cut all the meat from the skin, mince it small, and season it highly with pepper, nutmeg, and a clove, some parsley, lemon, thyme, sweet marjoram chopped, and a pounded onion, all well mixed, together with the well-beaten yolk of an egg; roll it up very tightly in the skin; tie it round, and bake it in an oven for two or three hours, according to the size of the mutton. Make a gravy of the bones and parings; season with an onion, pepper, and salt; strain and thicken it with flour and butter; add a tablespoonful each of vinegar, mushroom catsup, soy, and lemon pickle, and a teacupful of port wine; garnish with forcemeat balls made of grated bread, and part of the mince.

### Haggis

Wash and clean the heart and lights; parboil and mince them very small; add one pound of minced suet, two or three large onions minced, and two small handfuls of oatmeal; season highly with pepper and salt, and mix all well together; the bag being perfectly clean and sweet, put in the ingredients; press out the air, sew it up, and boil it for three hours.

### To Boil Marrow-bones

Saw them even at the bottom; butter and flour some bits of linen, and tie a piece over the top of each bone; boil them for an hour or two, take off the linen, and serve them with thin slices of dry toast cut into square bits. At table the marrow should be put upon the toast, and a little pepper and salt sprinkled over it.

### Beef or Mutton Baked with Potatoes

Boil some potatoes; peel and pound them in a mortar with one or two small onions; moisten them with milk and an egg beaten up; add a little salt and pepper. Season slices of beef or mutton chops with salt and pepper, and more onion, if the flavour is approved; rub the bottom of a pudding-dish with butter, and put in a layer of the mashed potatoes, which should be as thick as a batter, and then a layer of meat, and so on alternately, till the dish is filled, ending with potatoes. Bake in an oven for one hour.

## Olive Royals

Boil one pound of potatoes, and when nearly cold rub them perfectly smooth with four ounces of flour and one ounce of butter; knead all together till it becomes a paste; roll it out about a quarter of an inch thick, cut it into rounds, and lay upon one side any sort of cold roasted meat cut into thin small bits, and seasoned with pepper and salt; put a small bit of butter over the meat; wet the edges of the paste, and close it in the form of half-circles. Fry them in boiling fresh dripping till of a light brown colour; lay them before the fire, on the back of a sieve, to drain. Serve them with or without gravy in the dish. For a change, mince the meat, and season it as before directed. The potatoes should be very mealy.

## To Boil Ox-Cheek

Wash half a head very clean; let it lie in cold water for some hours; break the bone in two, taking care not to break the flesh; put it into a pot of boiling water, and let it boil from two to three hours; take out the bone. Serve it with boiled carrots and turnips. The liquor in which the head has been boiled may be strained, and made into barley broth.

## To Stew Ox-Cheek

Clean the head as before directed, and parboil it; take out the bone; stew it in part of the liquor in which it was boiled, thickened with a piece of butter mixed with flour, and browned. Cut into dice, or into any fancy shape, as many carrots and turnips as will fill a pint basin. Mince two or three onions, add the vegetables, and season with salt and pepper. Cover the pan closely, and stew it two hours. A little before serving, add a glassful of port wine.

## Dressed Ox-Cheek

Prepare it as directed for stewing; cut the meat into square pieces; make a sauce with a quart of good gravy, thickened with butter mixed with flour; season with salt and pepper, a little cayenne, and a tablespoonful of vinegar; put in the head, and simmer it till quite tender. A few minutes before serving add a little catsup or white wine. Forcemeat balls may be added.

### Potted Ox-Cheek

May be made of the meat that is left from any one of the above dishes. It is cut into small bits, and heated up with a little of the liquor in which the cheek was boiled, seasoned with pepper, salt, nutmeg, and a little vinegar, then put into a mould, and turned out when required for use. It is used for supper or luncheon, and is eaten with mustard and vinegar.

### Breasts of Mutton a la Ste. Menoult

Stew them with carrots, onions, and spices in gravy, and when done drain them and take out the bones; flatten the meat between two dishes, and when cold cut it into the form of cutlets or hearts; brush them with the beaten yolk of an egg; roll them in grated bread, then in clarified butter, and again in the grated bread. Bake them in an oven till of a fine brown colour, and serve them with an Italian or any other sauce.

### To Cure Mutton Ham

Cut a hind quarter of good mutton into the shape of a ham; pound one ounce of saltpetre, with one pound of coarse salt and a quarter of a pound of brown sugar; rub the ham well with this mixture, taking care to stuff the hole of the shank well with salt and sugar, and let it lie a fortnight, rubbing it well with the pickle every two or three days; then take it out and press it with a weight for one day; smoke it with sawdust for ten or fifteen days, or hang it to dry in the kitchen. If the ham is to be boiled soon after it has been smoked, soak it one hour; and if it has been smoked any length of time, it will require to be soaked several hours. Put it on in cold water and boil it gently for two hours. It is eaten cold at breakfast, luncheon, or supper. A mutton ham is sometimes cured with the above quantity of salt and sugar, with the addition of half an ounce of pepper, a quarter of an ounce of cloves, and one nutmeg.

### Meat or Birds in Jelly

Clean the meat or the bird or birds; fully roast, bake, or stew in the usual way.

Place the meat in the mould, or if birds, arrange them with their breasts downwards; fill the mould quite full with the jelly, recipe No. 329; set it to cool till the next day; then turn it on a dish, breasts upwards. Garnish the dish with curled parsley, and some of the jelly cut fine, and sprinkled about.

If the jelly be clear, it will form a handsome side-dish for dinner or supper.

## Pigeons in Savoury Jelly

Bone six pigeons; remove the heads and feet, stuff with sausagemeat, and roast; lay the pigeons in a mould with the breasts down; fill up the mould with jelly, recipe No. 329; and when cold, turn out. Garnish with parsley, and some of the jelly cut fine, and sprinkled round the dish.

## VEGETABLES

All vegetables should be boiled quickly, and, with the exception of spinach, in an open vessel, skimming them carefully.

Herbs of all sorts should be gathered when in flower, and on a dry day, and, being well cleaned from dust and dirt, tied up in small bunches and dried before the fire. They may then be kept in paper bags labelled; or rubbed to a powder, sifted, and put into bottles and labelled.

## To Boil Potatoes

Wash and pare them, throwing them into cold water as they are pared; put them into a saucepan, cover them with cold water, and throw in a little salt; cover the saucepan closely, and let them boil quickly for half an hour; pour off the water immediately, and set the pan by the side of the fire to dry the potatoes.

## To Boil Potatoes-Another Way

Wash them very clean, put them on in cold water, cover the saucepan, and let them boil quickly; as soon as the water boils pour it off, and cover them with cold water; add a little salt, and when the water boils pour it off instantly, when the potatoes will be sufficiently done; dry them, and take off the skins before serving.

## To Broil Boiled Potatoes

After boiling potatoes not quite sufficiently to send to table, put them on a gridiron over a clear fire, and turn them frequently till they are of a nice brown colour all over; serve them hot; take care they do not become too hard, as that spoils the flavour.

### To Brown Potatoes under Meat while Roasting

After being boiled, lay them on a dish, and place it in the dripping-pan; baste them now and then with a little of the meat dripping, and when one side is browned turn the other; they should all be of an equal colour.

### Potato Ribbons

Wash four or five large potatoes, scrape them, and cut them into thin strips round and round, keeping as nearly to one width as possible; throw them into cold water as they are cut, and then fry them of a light brown, in very hot or boiling beef dripping; strew over them a little salt and pepper, and before serving, drain them upon a dish turned up before the fire.

### To Boil Turnips

Wash, pare, and throw them into cold water; put them on in boiling water with a little salt, and boil them from two hours to two and a half; drain them in a colander, put them into a saucepan, and, mixing in a bit of butter, with a beater mash them very smoothly; add half a pint of milk, mix it well with the turnips, and make them quite hot before serving. If they are to be served plain, dish them as soon as the water is drained off.

### To Dress Young Turnips

Wash, peel, and boil them till tender in water with a little salt; serve them with melted butter poured over them. Or, They may be stewed in a pint of milk thickened with a bit of butter rolled in flour, and seasoned with salt and pepper, and served with the sauce.

### To Boil Spinach

Pick it very carefully, and wash it thoroughly two or three times in plenty of cold water; then put it on in boiling water with a little salt; let it boil nearly twenty minutes; put it into a colander, hold it under the water-cock, and let the water run on it for a minute; put it into a saucepan, beat it perfectly smooth with a beater or wooden spoon, add a bit of butter and three tablespoonfuls of cream, mix it well together, and make it hot before serving. When dished, it is scored in squares with the back of a knife.

## To Boil Spinach-Another Way

After being nicely picked and well washed, put it into a saucepan, with no more water than adheres to it; add a little salt; cover the pan closely, and boil it till tender, frequently shaking it; beat it quite smooth, adding butter and cream, and make it quite hot. Spinach may be served with poached eggs, or fried sausages laid on it.

When the spinach is bitter, it is preferable to boil it in water.

## To Boil Cauliflowers

Trim them neatly, and let them lie an hour or two in cold water; then rinse them in fresh cold water, and put them with a very little salt into boiling water; boil them twenty minutes, or half an hour if very large. They may be boiled in milk and water, and require to be skimmed with particular attention.

## To Boil French Beans

Cut off the stalk and string them; if not very young, cut them in four, or into very thin slices; put them into water as they are done, and put them on in boiling water, with a little salt, and let them boil for half an hour. If they are old they will require a longer time to boil. Melted butter in a sauce-tureen is served with them.

## To Boil Asparagus

Wash them well, scrape, and tie them up in small bundles; cut them all even at the bottom, and as they are done put them into cold water; put them on in boiling water, with a little salt, and let them boil twenty or twenty-five minutes; take them up, lay them upon a slice of toasted bread cut in four, and the crusts pared off, with the tops meeting in the middle of the dish, and cut off the strings.

## Asparagus a la Francais

Boil it, and chop small the heads and tender parts of the stalks, together with a boiled onion; add a little salt and pepper, and the beaten yolk of an egg; heat it up. Serve it on sippets of toasted bread, and pour over it a little melted batter.

## To Boil Brocoli

Wash it, cut off all the outside leaves and stalks, throw it into cold water as it is trimmed, put it on in boiling water with a little salt, and boil it for twenty-five minutes or half an hour. It is sometimes served upon bits of toasted bread, and a little melted butter poured round it.

### To Boil Artichokes

Cut off the stalks close to the bottom, wash them well, and let them lie for some hours in cold water; put them on in

boiling water with a little salt in it, cover the pan closely, and boil them an hour and a half. If they are old, and have not been freshly gathered, they will take a longer time to boil. Melted butter is served with them in a sauce-tureen.

### To Boil Young Green Cabbages

Wash and clean them well, put them on in boiling water with a little salt in it, and let them boil quickly from three-quarters to nearly an hour. Serve with melted butter.

### To Stew Cucumbers

Pare eight or ten large cucumbers, and cut them into thick slices; flour them well, and fry them in butter; then put them into a saucepan with a teacupful of gravy; season it highly with cayenne, salt, mushroom catsup, and a little port wine. Let them stew for an hour, and serve them hot.

### To Stew Cucumbers-Another Way

Pare the cucumbers, and let them lie in vinegar and water with a little salt in it; drain them, and put them into a saucepan with a pint of gravy, a slice of lean ham, an onion stuck with one or two cloves, and a bunch of parsley and thyme; let them stew, closely covered, till tender. Take out the cucumbers, strain and thicken the gravy with a piece of butter rolled in flour, boil it up, and pour it over the cucumbers.

### To Stew Mushrooms

Clean them as for pickling, and, after washing them, put them into a saucepan, with an anchovy, two cloves, some nutmeg sliced, mace, whole pepper, and salt; let them stew in their own liquor till tender.

In this way they will keep for some time, and when required to be dressed, pick out the spice, and to a dishful put two large tablespoonfuls of white wine; add part of their own liquor, and let them just boil; then stir in a bit of butter dredged with flour, and two tablespoonfuls of cream.

### To Stew Mushrooms-Another Way

For a good-sized dishful, take a pint of white stock; season it with salt, pepper, and a little lemon pickle; thicken it with a bit of butter rolled in flour; cleanse and peel the mushrooms, sprinkle them with a very little salt, boil them for three or four minutes, put them into the gravy when it is hot, and strew them for fifteen minutes.

### To Roast Onions

Roast them with the skins on in an oven, that they may brown equally. They are eaten with cold fresh butter, pepper, and salt.

### Onions, Plain Boiled

Peel them, and let them lie an hour in cold water, put them on in boiling milk and water; boil them till tender, and serve with melted butter poured over them.

### To Boil Carrots

Scrape, wash, and clean them; put them on in boiling water with some salt in it, and boil them from two to three hours. Very young carrots will require one hour.

### Carrots, Flemish Way

Prepare (after boiling) in the form of dice, balls, stars, crescents, &c., and stew with chopped parsley, young onions, salt and pepper, in plain melted butter, or good brown gravy.

### Green Peas Stewed

Put a quart of good peas into a stewpan, with a lettuce and small onion sliced small, but not any water; add a piece of butter the size of an orange, pepper and salt to taste, and stew gently for two hours. Beat up an egg, and stir into them (or a lump of butter will do as well); mint should be stewed (if it can be procured) with them, and ought to be chopped fine, and stirred in with some good gravy.

### To Boil Green Peas

After being shelled, wash them, drain them in a colander, put them on in plenty of boiling water, with a teaspoonful of salt, and one of pounded loaf sugar; boil them till they become tender, which, if young, will be in less than half an hour; if old, they will require more than an hour; drain them in a colander, and put them immediately into a dish with a slice of fresh butter in it. Some people think it an improvement to boil a small bunch of mint with the peas; it is then minced finely, and laid in small heaps at the end or sides of the dish. If peas are allowed to stand in the water after being boiled they lose their colour.

### To Stew Young Peas and Lettuce

Wash and make perfectly clean one or two heads of cabbage lettuce, pick off the outside leaves, and lay them for two hours in cold water with a little salt in it; then slice them, and put them into a saucepan, with a quart or three pints of peas, three tablespoonfuls of gravy, a bit of butter dredged with flour, some pepper and salt, and a teaspoonful of pounded loaf sugar. Let them stew, closely covered, till the peas are soft.

### Peas for a Second-course Dish, a la Francais

Put a quart of fine green peas, together with a bit of butter the size of a walnut, into as much warm water as will cover them, in which let them stand for eight or ten minutes. Strain off the water, put them into a saucepan, cover it, stir them frequently, and when a little tender add a bunch of parsley and a young onion, nearly a dessertspoonful of loaf sugar, and an ounce of butter mixed with a teaspoonful of flour; keep stirring them now and then till the peas be tender, and add, if they become too thick, a tablespoonful of hot water. Before serving, take out the onion and bunch of parsley.

### To Steam Peas

Shell and close-pack the peas securely in a large quantity of lettuce-salad leaves; put the package into a stewpan over a moderate fire for the ordinary time required to boil peas, say half an hour, when they will be ready.

### Vegetable Mash

Take boiled potatoes, cauliflower, carrots, turnips, and green peas; mash down the potatoes with plenty of butter, pepper, and salt; mince small the cauliflower, carrots, and turnips, and add them with the peas to the mashed potatoes; mix them all well together, and serve up hot.

## PASTRY, PUDDINGS, SWEETMEATS, ETC.

### Pastry for Pies and Tarts

To every three ounces of flour take one ounce of soojee, two ounces of beef suet, and a little salt; pick and clean the suet, pound it in a mortar, and make a flat circular cake of it; make a dough of the flour and soojee, knead it well, divide it into two equal parts, and make them into two flat circular cakes quite as large as the suet cake; roll the three together, placing the suet cake between the two flour cakes; double the whole up twice, and roll it out again, when it will be ready for use.

### Pastry for Friar Tuck's Mock Venison Pastry Pie

Take of veal one pound, and of udder one pound; pick, clean, and wash them; chop, mince, and pound them in a mortar; season with salt and white pepper; fix the mixture with the yolk and white of an egg well beaten up; pass it through a sieve, rejecting all that will not pass; and form it into a flat circular cake.

Make a dough of two pounds and a half of flour and half a pound of soojee; add a little salt, and knead it well; then form two cakes of the dough; place the veal and udder cake between, and roll out the three very carefully; double up the whole and roll it out again, when it will be ready.

The pie-dish should be lined thickly with the pastry, and, although a single layer should cover the top of the pie, the sides and edges of the dish should be built up high with it; a double layer of the crust is not interdicted to cover the top of the pie if it will not interfere with raising it up.

### Custard

Take a seer of milk and a stick of cinnamon, and boil down to half the quantity; add sugar to taste; beat up quickly the yolks of four eggs, and add them gradually to the milk, stirring it continually; after a while thicken with a tablespoonful of rice flour; take it off the fire, and flavour with rose water, orange-flower water, or vanilla.

### Orange Custard

Boil very tender the rind of half an orange, and beat it in a mortar until it is very fine; put it to a spoonful of the best brandy, the juice of an orange, four ounces of loaf sugar, and the yolk of four eggs; beat them altogether for ten minutes, and then pour in by degrees a pint of boiling milk; beat them until cold; then put them in custard-cups into a dish of hot water; let them stand till they are set; then take them out, and stick preserved orange-peel on the top. This forms a fine-flavoured dish, and may be served up hot or cold.

### Chocolate Custard

Rasp three ounces of fine Spanish chocolate, which has the vanilla flavour; make a paste of it with the smallest possible quantity of water; put two pints of pure milk over the fire, and let it boil; then add powdered loaf sugar to your taste, and a /little/ salt; meanwhile, beat up the chocolate with some of the milk as it boils, and mix it well; pour it into the boiling milk, which you must keep in motion; add the yolks of eight eggs well beaten up; keep stirring in, or rather milling the mixture, until of sufficient consistency; when cool enough put the custards into glasses.

### Almond Custard

Blanch and pound, with two tablespoonfuls of orange-flower water, a quarter of a pound of almonds; add rather more than a pint of milk, thickened with a teaspoonful of corn-flour, and the well-beaten-up yolks of six eggs; sweeten to taste with pounded loaf sugar, and stir it over a slow fire till it thickens, but do not allow it to boil. Serve up in glass custard-cups.

### Princess Royal Custard

Beat up in a large deep bowl the yolks of eight fresh eggs; dredge into it while beating up a dessertspoonful of corn-flour; sweeten to taste with the best pounded loaf sugar; add to it a quarter of a pound of Jordan almonds well bruised in a marble mortar; pour the mixture into a clean newly-tinned copper pan; stir into it a seer of good cold milk; have a brisk flaming fire ready. Put the pan on the fire; never cease stirring it, keeping the spoon as much as possible in the centre of the pan; reduce the flame after it has boiled for fifteen minutes, and continue to boil for a few minutes longer, until the custard is of the consistency required.

### Rose-bloom Custard

This is made in every respect like the foregoing, adding some bruised almonds, and a little rose-bloom to tint the custard. The froth of the white of the eggs is also tinted with a few drops of the rose-bloom.

## Blanc Mange

Boil, till dissolved, three-fourths of an ounce of isinglass in as much water as will cover it; when lukewarm, add to it gradually a quart of good rich milk, with a stick of cinnamon, some lemon-peel, and a few bitter almonds well pounded; sweeten to taste; boil for five or six minutes, stirring it all the while; strain through muslin into moulds, and place in a pan of cold water to congeal.

## Blanc Mange-Another Way

Blanch and pound with a little rose-water two ounces of sweet and six bitter almonds; dissolve three-fourths of an ounce of isinglass in a little water; when nearly cool, mix it into a quart of good rich milk; mix in the almonds the peel of a small lemon and a stick of cinnamon; sweeten to taste with good clean sugar; let it stand for two or three hours; then put it into a pan, and let it boil for six or eight minutes, stirring it constantly; strain through muslin, and keep stirring it until nearly cold; then pour it into moulds.

## Rice Blanc Mange

Mix to a stiff smooth paste four tablespoonfuls of finely-sifted ground rice-flour, with a little cold milk; then stir it into a quart of boiling milk, in which had been dissolved one-eighth of an ounce of isinglass, a stick of cinnamon, and the peel of half a small lemon; sweeten to taste; boil it from ten to fifteen minutes, stirring it carefully all the while; remove it from the fire, and mix into it briskly a tablespoonful of pounded almonds, and pour it while scalding hot into moulds previously dipped in /cold/ water.

## Corn-flour Blanc Mange

The above recipe will answer, except that the quantity of corn-flour must be in the proportion of two tablespoonfuls to every quart of milk.

## Christmas Plum Pudding (Indian Way)

This pudding may be made a few days before it is required for the table.

Take of cleaned and picked raisins one pound and a half, currants half a pound, finely-grated bread-crumbs three-quarters of a pound, finely-sliced mixed peel half a pound, finely-minced suet three-quarters of a pound, and sugar three-quarters of a pound. Mix all these well together with half a teaspoonful of finely-powdered mixed spices, say cinnamon, nutmeg, and mace; then moisten the mixture with half a pound of butter free of water, twelve eggs well beaten, and a wineglassful of brandy, stirring it well the whole time, that the ingredients may be thoroughly mixed.

Butter a large piece of cloth or napkin; dredge it well with flour; put the mixture into it, and tie it down tightly; after boiling it steadily for seven hours take it out of the boiler and hang it up immediately, until the day it is intended to be eaten, when it should be boiled again for fully two hours, care being taken that the water is boiling before the pudding is put into it. Then turn it out of the towel, and serve up with brandy sauce.

### Bombay Pudding

Take two pounds or one seer of soojee, half roast it, then boil it in water until it becomes very thick; butter a soup-plate or any other dish of about the same depth; pour the boiled soojee into it; when it has cooled and congealed, cut it into eight or more cakes; rub the cakes over with the yolk of an egg, dredge with finely-sifted flour, and fry in ghee until they acquire a rich brown colour. Arrange them in a dish, and pour over them a thick syrup flavoured with lemon-juice.

### Bombay Pudding-Another Way

Make a good sweet custard and set it aside; rasp fine a cocoanut, and fry it in a little butter with grated nutmeg; pour into it gradually a wineglassful of brandy, stirring it all the time; have a pudding-dish lined with a good puff paste; pour the fried cocoanut gradually into the custard, stirring it well all the while; fill the pudding-dish with the mixture, and bake it in a gentle oven for fifteen to twenty minutes, or until the pudding is cooked.

### Cocoanut Rice Pudding

Soak a breakfastcupful of fine rice in water until quite soft; scoop out the contents of a hard cocoanut; extract all the milk with a little boiling-hot water, then boil the rice in it, sweeten it to taste with some date jagree or treacle, and put in a few grains of aniseed. Pour the mixture into a buttered pudding-dish and bake it slightly.

### Indian Lemon Pudding

Take four chittacks or eight ounces of butter free of water, six chittacks or twelve ounces of white sugar, twelve fresh eggs, four wineglassfuls of lemon or lime juice, and four tablespoonfuls of finely-grated bread-crumbs. Mix the butter and the sugar, add the yolks of the eggs, then the lime-juice and bread-crumbs, and when the oven is ready add the whites of the eggs well beaten up, put the whole into a buttered pudding-dish, and bake it immediately.

### Marmalade Pudding

This pudding requires care in mixing the ingredients thoroughly together, but it proves so excellent when eaten, either cold or hot, that it fully repays the trouble of preparation. Shred six ounces of fresh beef suet, and chop it up fine; mix it with two ounces of moist sugar, a quarter of a pound of well-grated bread-crumbs, and then stir in half a pint of new milk; when these are all mixed, add the well-beaten yolks of three eggs, whisk all together for a quarter of an hour, and set it to stand on a cold stone for an hour.

Butter a pudding-dish or mould thickly, place a layer of the above mixture in it, then a layer of marmalade, another layer of mixture, and so on alternately until the mixture is exhausted. For the above quantity about one pound of marmalade will be required. Whisk the whites of the eggs with a little loaf sugar and orange-flower water, place the froth at the top of the pudding, and bake for an hour and a half in a moderate oven.

### Custard Pudding

Mix with a pint of cream or milk six well-beaten eggs, two tablespoonfuls of finely-sifted flour, half a small nutmeg grated, or an equal quantity of pounded cinnamon, a tablespoonful of pounded loaf sugar, and a little salt; put it into a cloth or buttered basin, that will exactly hold it, and boil it for half an hour. Serve with wine sauce.

### Macaroni

Take the yolks and white of two fresh eggs, and as much finely-sifted flour (English or American preferable to country) as will make a good dough of the consistency of dough for pie-crusts without the addition of any water; roll it out to its full extent on a large board to about the thickness of an eight-anna piece; then cut it up into small squares, diamonds, or circles, or into any shape or design you please, which must be done quickly, as within an hour of its being rolled out the pastry will harden. It may be used immediately, or in the winter it may be kept good for a few days.

N.B.—If pipe macaroni be required, cut the macaroni in ribbons of the required width, dredge some flour over it, and put it lengthways over glass pipes, joining the two cut ends with the aid of a little raw egg; and draw the pipes out as the pastry hardens round them. For pipe macaroni, the pastry should be rolled finer.

## Tart and Pie Crusts of Soojee

To one seer and a quarter of soojee add half a seer of suet and a teaspoonful of salt. Thoroughly clean the suet, remove all the skin and other objectionable particles, chop, mince, and pound fine in a mortar. Damp the soojee for half an hour before kneading it, then knead it with the suet and a little of the yeast, recipe No. 283; divide it into parts, dredge it with flour, and roll in layers; repeat the operation two or three times, and the pastry when baked will be light and flaky. Half a seer of flour will be required for dredging and rolling.

## Chappatee or Hand-Bread

The native hand-bread is made simply of wheat-flour and water; the addition of a little salt would be an improvement. Make a good dough of flour and water, take a piece about the size of an egg, roll it out to the circumference of a half-plate, and bake it over an iron or earthen plate.

## Dalpooree

Prepare a dal chur churree, recipe No. 93; put it into a marble mortar, and reduce it to a fine paste.

Prepare an ordinary pie pastry; take two pieces of the prepared dough, each of the size of a walnut; shape them into two small bowls; take as much of the dal paste as will make a ball the size of a walnut; put it into one of the bowls of dough, and cover it over with the other bowl, and then roll out the whole very carefully to the size of a dinner-plate, and fry in ghee of a very light yellow colour. The lighter and thinner dalpoorees can be made the better. They should be eaten hot.

## Dal Pittas

Prepare an ordinary pie-crust, and the dal chur churree, recipe No. 93; roll out the pastry, cut into circles of the size of saucers, put into them a tablespoonful of the dal, and close them; fry in ghee of a light brown colour. They should be eaten hot.

## Prawn Doopiaja Pittas

The same as the above, enclosing in the pastry a tablespoonful of the prawn doopiaja, recipe No. 69; fry in ghee.

## Prawn Doopiaja Loaf

Pare away very finely all the outer brown crust of the bread, without injuring the inner crust; cut out of the top of the loaf a small square sufficiently large to extract from within all the crumb, leaving the shell complete; then fill the loaf up to the top either with some prawn doopiaja minced, or with the prawn cofta curry, No. 37, and as much gravy as it will take; replace the square bit at the top, bake it to a light brown, and serve up hot.

## Fowl Doopiaja Loaf

Is made in the same way as the prawn loaf, the difference being that the shell of the bread is stuffed with either a fowl doopiaja, recipe No. 23, or with the chicken cofta curry, recipe No. 34; all the bones of the fowl will require to be removed before the bread is stuffed with the curry.

## Falooree

Take of the finely-sifted flour of the chunna ka dal, which has been previously parched, one seer; six large Patna onions finely sliced and chopped; eight fresh green chilies sliced very fine; a tablespoonful each of finely-chopped soa mattee, saug, and parsley; a dessertspoonful of salt and a teaspoonful of finely-ground green ginger. Put the seer of dal-flour into a large deep pan, and mix into it all the above condiments; then keep adding to it water, very gradually and in small quantities at a time, mixing it briskly the whole while, until it is of a consistency that if poured on a plate from a spoon it will incline to a pyramid, or if dropped into a glass of water will not readily dissolve, but drop to the bottom /en masse/. In this state the mixture will be ready to fry.

Take half a seer of the best mustard oil; put it into a deep frying-pan with some fine slices of lemon-peel, and fry it or cook it thoroughly; remove three-fourths of the cooked oil from the frying-pan, and into the remainder, while boiling and bubbling, with a tablespoon pour in the preparation in the shape of rocks, and allow to brown, turning them over so that top and bottom may be of the same colour. As the oil is being expended clear the pan of all particles which may accumulate, pour in some more of the ready-cooked oil, and continue to fry until all the mixture is fried. They should be eaten hot.

### Cocoanut Pittas

Scrape finely a cocoanut, brown it with some jagree and a few grains of the black cardamom seed, and set it aside; then prepare a pastry of finely-sifted rice-flour (it must be kneaded with boiling-hot water, and will not roll out); take as much as the size of a duck's egg, and press it out flat in the palm of your hand to the size of a large saucer; put a tablespoonful of the fried cocoanut into it, and close it up in a half-moon shape, with the help of a little water. Have a wide-mouthed large earthen pot of boiling water; stretch and tie over its mouth a napkin, and steam the pittas or cakes over them; they will be ready in half an hour, and may be eaten hot or cold.

### Plantain Fritters

Prepare a batter of twelve ripe plantains, four tablespoonfuls of finely-sifted flour, half a cupful of milk, sugar to taste, and cardamom and caraway seeds, with a couple of eggs beaten up; mix the whole well together, and make into small cakes by pouring a tablespoonful at a time of the mixture into melted ghee; fry them on both sides to a good brown colour, and serve up hot.

### Fried Plantains

Slice or divide very ripe plantains lengthways into two; brush them slightly with the yolk of an egg; dredge with flour, and fry in melted ghee. Serve up hot, sprinkled with crushed crystallized sugar.

### Bibinca Dosee, or Portuguese Cocoanut Pudding

Extract a cupful of milk from two cocoanuts, and set it aside. Make a syrup of three-quarters of a pound of sugar; mix into the syrup half a pound of rice-flour finely sifted, and the cocoanut milk, which boil over a good fire, stirring the whole while until it thickens; pour it into a buttered pudding-dish, and bake it of a rich light-brown colour.

### Bole Comadree, or Portuguese Cocoanut Pudding with Jagree

Extract a cupful of milk from two cocoanuts, and set it aside. Make a syrup of half a pound of sugar; mix into it half a pound of finely-sifted rice-flour, and set aside; fry with the yolk of an egg all the scrapings of the two cocoanuts, half a pound of jagree, and some grains of aniseed; then mix the whole thoroughly together, and after the oven is well heated, and ready to receive the pudding, pour the mixture into a well-buttered pudding dish, and bake over a slow fire until it is perfectly set.

### Goolgoola, or Fritters

Take half a seer or one pound each of flour, sugar, and milk, half a dozen small sticks of cinnamon, a little yeast, and half a seer of ghee; mix the flour with the yeast and a little milk; add water sufficient to bring it to a thick consistency; then put into it gradually the sugar and the remainder of the milk, and place it on the fire, adding the cinnamon; keep stirring it with a large spoon until it is again reduced to a thick consistency; remove it from the fire, and when it has cooled make it up into small balls, and fry them in ghee.

### Goolgoola, or Fritters-Another Way (as usually served on the tea-table)

Take two chittacks or four ounces of soojee, four eggs well beaten up and four chittacks or eight ounces of milk; mix the soojee and eggs, beating them well together, and gradually add the milk. Melt down three chittacks or six ounces of ghee in a small but deep pan; pour into the boiling ghee in one spot the mixture, a dessertspoonful at a time, and fry until of a rich brown colour. Serve up hot, sprinkled with crushed crystallized sugar.

### Cajure

Mix one seer of soojee with four tablespoonfuls of ghee; add half a seer of sugar; mix well together; then pour in gradually a quarter of a seer of milk, and last of all as much flour as will make a good dough; let it be well kneaded, and then allowed to stand for two or three hours.

Have some ghee melted; take the dough of the size of walnuts, shape them like shells and fry them in the melted ghee until they acquire a rich brown colour.

### Hulluah

Steep half a seer of soojee in one seer of water for twelve hours, or, if the hulluah be made in the winter, let it soak for eighteen hours; it will then be the "milk of soojee," which strain through a coarse duster, rejecting only such impurities as remain unstrained; add to the milk half a seer of sugar, and boil it, stirring it all the time, and as it thickens add three chittacks or six ounces of ghee, warmed with a few white cardamoms and a few small sticks of cinnamon; continue stirring it from first to last until the whole is well mixed together, and the hulluah finally taken out of the pan; while warm put it into shapes or moulds.

### Hulluah-Another Way

Take half a seer of soojee, ghee, sugar, almonds, and raisins, and a few white cardamoms and sticks of cinnamon. Make a syrup of the sugar, and set it aside. Roast the soojee, or brown it, and set it aside. Melt the ghee, and fry the soojee with the spices in it, after which put in the almonds and raisins, stirring it well all the time; last of all add the syrup, and continue to cook and stir it until it thickens; then remove into moulds or shapes while hot.

### Swiss Cakes

Take butter, flour, and sugar, of each the weight of four eggs; beat the yolks with the sugar and some grated lemon-peel, or ten drops of essence of lemon, and one large teaspoonful of rose-water, or orange-flower water if preferred; add the butter just melted, and slowly shake in the flour, beating it until well mixed; beat the whites of the eggs to a froth, mix the whole together, and beat on for a few minutes after the whites are added. Butter a tin, and bake the cake half an hour.

### Queen Cakes

Prepare eight ounces of fresh butter beaten to a cream, six ounces of pounded and sifted loaf sugar, half a pound of dried and sifted flour, the same quantity of cleaned and dried currants, four well-beaten eggs, a little grated nutmeg and pounded cinnamon, and a few pounded bitter almonds; then add the sugar to the butter, put in the eggs by degrees, after that the flour and the other ingredients; beat all well together for half an hour, and put it into small buttered tins, nearly filling them, and strew over the top finely-powdered loaf sugar. Bake them in a pretty brisk oven.

## Shrewsbury Cakes

Mix with half a pound of fresh butter, washed in rose-water and beaten to a cream, the same quantity of dried and sifted flour, seven ounces of pounded and sifted loaf sugar, half an ounce of caraway-seeds, and two well-beaten eggs; make them into a paste, roll it thin, cut it into round cakes, prick them, and bake them upon floured tins.

## Shrewsbury Cakes-Another Way

Rub into a pound of dried and sifted flour half a pound of fresh butter, seven ounces of sifted loaf sugar, the same quantity of cleaned and dried currants, and a little grated nutmeg; make it into a paste with a little water and two tablespoonfuls of rose or orange-flower water; roll it out, and cut it into round cakes; prick them, and bake them upon tins dusted with flour.

## Shortbread

For two pounds of sifted flour, allow one pound of butter, a quarter of a pound of candied orange and lemon-peel, a quarter of a pound each of pounded loaf sugar, blanched sweet almonds, and caraway comfits; cut the lemon, the orange-peel, and almonds into small thin bits, and mix them with a pound and a half of the flour, a few of the caraway comfits, and the sugar; melt the butter, and when cool, pour it into the flour, at the same time mixing it quickly with the hands; form it into a large round nearly an inch thick, using the remainder of the flour to make it up with; cut it into four, and with the finger and thumb pinch each bit nearly all round the edge; prick them with a fork, and strew the rest of the caraway comfits over the top. Put the pieces upon white paper dusted with flour, and then upon tins. Bake them in a moderate oven.

## Scotch Shortbread

Warm before the fire two pounds of flour and one pound of butter free of water; rub the butter, with twelve ounces of sugar, into the flour with the hand and make it into a stiff paste with four eggs, well beaten; the rolling-out to the required thickness must be done with as little use of the rolling-pin as possible; either take small pieces, and roll them into oblong cakes, or roll out a large piece and cut it into squares or rounds; prick a pattern round the edge of each cake with the back of a knife, and arrange slices of candied peel, caraway-seeds, and caraway comfits in a pattern. They will take about twenty minutes to bake, and the oven itself should not be too quick. The mixing of flour, sugar, and butter, and afterwards of the eggs, must be done very thoroughly and smoothly.

### Scotch Shortbread-Another Way

Take two pounds of flour, one pound of butter, four eggs, and twelve ounces of loaf sugar powdered very finely; rub the butter and sugar into the flour with the hand, and by means of the eggs convert it into a stiff paste; roll it out half an inch thick, and cut into square or round cakes; pinch up the edges to the height of about an inch, and on the top of each cake place some slices of candied peel and some large caraway comfits, pressed down so as to imbed about half of each in the cake. Bake in a warm oven upon iron plates.

### Gingerbread Nuts

Take three pounds of flour, a pound of sugar, three pounds and a half of treacle, half an ounce of caraway-seeds, half an ounce of allspice, two ounces of butter, half an ounce of candied lemon-peel, three ounces of ground ginger, half an ounce of coriander, the yolks of three eggs, and a wineglassful of brandy; work the butter to a cream, then the eggs, spice, and brandy, then flour, sugar, and then hot treacle; if not stiff enough, a little more flour must be added in rolling out, but the less the better.

### Gingerbread Nuts-Another Way

Take two pounds of flour, one pound and a quarter of treacle, half a pound of sugar, two ounces of ginger, three-quarters of a pound of butter (melted), and a small quantity of cayenne pepper; mix all together and roll out to about the thickness of half an inch, or not quite so much; cut into cakes, and bake in a moderate oven.

### Ginger Cakes

In two pounds of flour well mix three-quarters of a pound

of good moist sugar and one ounce of the best Jamaica ginger; have ready three-quarters of a pound of lard melted, and four eggs well beaten; mix the lard and eggs together and stir into the flour, which will form a paste; roll out into thin cakes and bake in a moderately heated oven.

Lemon biscuits may be made the same way, substituting essence of lemon instead of ginger.

### Gingerbread Spiced

Take three-quarters of a pound of treacle, one egg, four ounces of moist sugar, an ounce of powdered ginger, a quarter of an ounce each of mace, cloves, allspice, and nutmeg powdered, a pound of oiled butter, and sufficient flour to make a stiff paste; mix well, and make into thick pieces, which should be brushed over the top with white of egg and baked for an hour in a moderate oven.

### American Gingerbread

Take half a pound of fresh butter melted, a pound and a half of dried and sifted flour, the same quantity of brown sugar, a quarter of a pound of pounded ginger, nine eggs, the yolks and whites separately beaten, one glass of rose-water, and one of white wine; mix all well together, and beat for an hour; then with a spoon spread it over flat tin pans, about the thickness of a penny-piece; bake it of a light brown, and while warm cut it into oblong pieces, and place them on end till cool, when they will be very crisp.

### Rich Gingerbread Cakes

To one pound of dried and sifted flour allow half a pound of pounded loaf sugar, three-quarters of a pound of fresh butter washed in rose-water, one pound of treacle, one nutmeg grated, the weight of a nutmeg of pounded mace, and as much of pounded cinnamon, one ounce of pounded ginger, one and a half of candied orange and lemon-peel cut small, half an ounce of blanched sweet almonds cut into long thin bits, and two well-beaten eggs. Melt the butter with the treacle, and when nearly cold stir in the eggs and the rest of the ingredients; mix all well together, make it into round cakes, and bake them upon tins.

### Indian Gingerbread

Take twelve ounces of pounded loaf sugar, a quarter of a pound of fresh butter, one pound of dried flour, two ounces of pounded ginger, and a quarter of an ounce each of cloves and cinnamon. Mix the ginger and the spice with the flour; put the sugar and a small teacupful of water into a saucepan; when it is dissolved add the butter, and as soon as it is melted mix it with the flour and other things; work it up, form the paste into cakes or nuts, and bake them upon tins.

### Oatmeal Gingerbread

Gingerbread made with oatmeal instead of flour, besides being nice, is a very useful aperient for children.

### Excellent Cheesecakes, known at Richmond as "Maids of Honour"

Take half a pound of curd free from the whey; add to it six ounces of butter, four yolks of eggs, and sugar and nutmeg to the taste; mix all the ingredients well; line patty-pans with a puff paste, fill them with the mixture, and bake in a quick oven. The cheesecakes may be flavoured with lemon for a variety, and, as a further variety, currants and raisins may be introduced.

### Cocoanut Cheesecakes

Grate a good-sized nut very fine, and add to it four or five spoonfuls of rich syrup and one spoonful of rose-water; set it over a few coals, and keep stirring till it is mixed; then take it off the fire and let it cool; next mix the yolks of two eggs well with it, and bake in small paps in the shape of cheesecakes. The pastry for the pans must be made with flour and yokes of eggs, rolled as thin as possible; wet the tops of the cakes with rose-water; sift some refined sugar over them, and bake them in an oven at a gentle heat.

### Buns

Mix together one pound of flour, six ounces of butter, two teaspoonfuls of baking powder, a quarter of a pound of sugar, one egg, nearly a quarter of a pint of milk, and a few drops of essence of lemon. Bake immediately. The above quantities will make twenty-four buns; for variety, currants or raisins may be added.

### Rout Cakes

To one pound of ground almonds add one pound of powdered sugar; mix them together with yolks of eggs to a stiff, yet flexible paste; then form it into small biscuits in the shape of coronets, shells, filberts, birds' nest, rings, or any other fancy shapes; let them remain five or six hours, or all night, upon the baking-tin in a warm oven.

## French Pancakes

Beat separately the yolks and whites of seven eggs; beat with the yolks four tablespoonfuls of pounded loaf sugar, the same quantity of flour, one pint of cream or milk, the grated peel and juice of one lemon, and two tablespoonfuls of rose-water; add the beaten whites the last thing. Allow three tablespoonfuls to each pancake.

## Common Pancakes

With nearly half a pound of flour mix five well-beaten eggs, and then add, by degrees, a quart of good milk; fry them in fresh lard, and serve them with pounded loaf sugar strewed between each.

## Indian Pancakes

Add to three well-beaten eggs a pint of new milk, three tablespoonfuls of flour, some sugar, and a little pounded cinnamon; mix all well together, and fry in butter; brown the upper side for a minute before the fire; serve it, cut into four, with pounded sugar strewed over it.

## Pink Pancakes

These are rarely seen at an English table, although they form a very pleasing variety. Boil a large red beetroot until it is very tender; then peel it, cut it into thin slices, pound it to a pulp in a marble mortar, and strain through muslin; add the yolks of five eggs, two tablespoonfuls of flour, four of cream, plenty of pounded loaf sugar, half a nutmeg grated, and a wineglassful of brandy; rub the whole into a batter, and fry the pancakes with melted butter, ghee, or lard; serve them up hot, garnished with green candied sweetmeats.

## Mango Fool

Take six green mangoes; remove every particle of the green peel, cut them into four, and steep them in clean water; throw the stones away; boil the fruit perfectly tender, pulp and pass it through a sieve, sweeten to your taste, and add to it very gradually, stirring all the while, as much good pure milk as will reduce it to the consistency of custard. It should be eaten on the day it is made.

### Mango Fool-Another Way

Boil to a pulp some green mangoes without peel or stones; pass through a sieve, and sweeten to taste; then mix into it very gradually some cold milk, which has been previously boiled; keep stirring until it has acquired the thickness of an ordinary cream custard; fill into glass cups, and grate a little cinnamon or nutmeg over them.

### Pink Mango Fool

The pink mango fool is produced by the introduction of beetroot boiled very tender, bruised down, strained through muslin, and added to the pulp of the mango, and forms an agreeable variety.

### Mincemeat

Ingredients-Three large lemons, three large apples, one pound of stoned raisins, one pound of currants, one pound of suet, two pounds of moist sugar, one pound of sliced candied orange-peel, one ounce of sliced candied citron, the same quantity of lemon-peel, one teacupful of brandy, and two tablespoonfuls of orange marmalade.

Grate the rinds of the lemons, squeeze out the juice, strain it, and boil the remainder of the lemons until tender enough to pulp or chop very finely; then add to this pulp the apples, which should be baked, and their skins and cores removed; put in the remaining ingredients one by one, and as they are added mix everything thoroughly together. Put the mincemeat into a stone jar with a closely-fitting lid, and in a fortnight it will be ready for use. This should be made the first or second week in December.

### Mincemeat-Another Way

Take seven pounds of currants well picked and cleaned; of finely chopped suet beef, the lean of sirloin of beef minced raw, and citron, lemon, and orange peel cut small, each half a pound; two pounds of fine moist sugar, an ounce of mixed spice, and the rinds of four lemons and four oranges; mix well, and put in a deep pan. Mix a bottle of brandy and white wine and the juice of the four lemons and oranges; pour half over, and press down tight with the hand; then add the other half and cover closely. It may be made one year, to use the next.

### Ornaments for Custards or Creams

Whisk for an hour the whites of two eggs, together with two tablespoonfuls of some syrup or thin jelly; lay it in any form upon a custard or cream, piled up to imitate rock, or it may be served in a dish with cream round it. The ornament may be coloured, if desired, with cochineal, saffron, spinach, &c., as directed in the following recipe.

### Colouring Mixtures

Yellow.—Into a four-ounce phial put half a drachm of saffron and two ounces of spirits of wine of the strength of sixty-two degrees over proof. Let it stand until the spirit is tinted of a deep yellow; then strain it for use.

Red.—This is produced by infusing during a fortnight two ounces of red sandal-wood in a pint of spirits of wine. It at the expiration of that time the colour should not be dark enough, a pinch of subcarbonate of soda will give it the required tint.

Pink.—Dissolve half an ounce of cochineal in a sufficient quantity of spirits of wine.

Green.—Put a handful of well-cleansed vine-leaves or spinach into a decanter, fill with spirits of wine, and let it stand in the sun for ten or twelve days; strain when the wine has become of a bright green.

N.B.—The above colouring matters are only adapted for tinting liqueurs, wines, lemonades, and essences.

### Frost or Icing for Cakes

Beat till very light the whites of four eggs, and add gradually three quarters of a pound of double-refined sugar, pounded and sifted through a lawn sieve; mix in the juice of half a lemon; beat it till very light and white; place the cake before the fire, pour over it the icing, and smooth over the top and sides with the back of a spoon.

### Frost or Icing for Cakes-Another Way

Beat to a stiff froth the whites of three new-laid eggs, and add to them one pound of sifted white sugar; flour the cake, and then wipe it off; apply the icing by means of a knife smoothly; then bake in a slow oven.

## Coloured Icings

Pink icing should be made by adding cochineal syrup; blue, with indigo; yellow, with saffron or gamboge; green, with spinach syrup or sap green; brown, with chocolate.

## Fine Icing for Tarts and Puffs

Pound and sift four ounces of refined loaf sugar; beat up the white of an egg, and by degrees add it to the sugar till it looks white and is thick.

When the tarts are baked, lay the icing over the top with a brush or feather, and then return them to the oven to harden, but take care that they do not become brown.

## Raspberry Iced Cream

Mix a tablespoonful of pounded loaf sugar, two tablespoonfuls of raspberry jelly or jam, and a little cochineal to heighten the colour, with the juice of a large lemon; strain, and put into the freezing-pot; cover it closely and place it in a bucket which has a small hole near the bottom, and a spigot to let the water run off, with plenty of ice broken small, and mixed with three or four handfuls of coarse salt; press the ice closely round the freezing-pot, turn it round and round for about ten minutes, take off the cover, and remove with a spoon the frozen cream to the centre; cover it again, and turn it till all be equally iced.

Serve it in china ice-pails in block, or put it into moulds, cover them securely, and replace them in the bucket, with ice and salt as before, for an hour or more; dip the moulds into cold water before turning out, and serve immediately. Water ices are made in this way, substituting water for cream.

## Apricot Iced Cream

Mix a tablespoonful of pounded loaf sugar with two of apricot jam, the juice of a lemon, and half an ounce of blanched bitter almonds pounded with a little rose-water; add a pint of cream, stir all well together before putting it into the freezing-pot, and freeze it as directed above.

## Mille Fruit Iced Cream

Strain the juice of three lemons, and grate the peel of one; mince finely a dessertspoonful each of orange marmalade, dried cherries, and preserved angelica; add to these half a pint of syrup, and mix the whole with a pint and a half of cream, or a pint of water, and then drop in here and there a few drops of the prepared cochineal. Put it into a mould, and freeze as above directed.

## Orange-water Iced

Mix with a pint of water the strained juice of three oranges and one lemon, also the grated peel of one orange; sweeten it well with syrup, and freeze it.

## Juice of Fruit Iced

Press through a sieve the juice of a pint of currants or raspberries, or other fruit preserved for tarts; add to it four or five ounces of pounded loaf sugar, a little lemon-juice, and a pint of cream. It may be whisked previous to freezing, and a mixture of the juice in which the fruit was preserved may be used.

## Orange Iced Cream

Boil down a seer and a half of milk to half the quantity with some isinglass and a quarter of a seer or half a pound of sugar; strain through a sieve, and when perfectly cool add the juice of twelve oranges. Mix well, put into freezing-pots with two seers or four pounds of raw rice and some salt, and freeze as above.

## Bael Sherbet

Take a perfectly ripe sweet bael, and scoop out the whole contents into a bowl; make a paste of it with a little water; then add sugar to taste, and as much water as will bring it to the consistency of good honey; then pass it through a fine sieve, leaving all the fibres and seeds behind; it is a most delicious drink, and if taken early in the morning in rather a liquid state—say of the consistency of porter—serves as a most effective aperient in a natural and healthy form; but if taken of the consistency of thick pea or potato soup, it has a directly contrary effect, and as such is invaluable in all cases of relaxed bowels.

## Mallie, or Cream as prepared by the Natives

Boil down over a slow fire milk to less than half its original quantity, and when cold it will be of the strength and consistency of a well-made blanc mange.

N.B.—The best Indian sweetmeats are made of mallie.

### Tyre or Dhye

Warm some milk without boiling it; stir into it a little stale butter about the size of a large pea; put the vessel in a warm place well covered over, and in the course of eight or ten hours the tyre will be ready.

### Yeast

Boil one pound of good flour, a quarter of a pound of brown sugar, and a little salt in two gallons of water for one hour; when milk-warm, bottle it close; it will be fit to use in twenty-four hours. One pint of this will make eighteen pounds of bread.

### Yeast-Another Way

Take two pounds of soojee or flour, a quarter of a pound of brown sugar or suckur, and half a drachm of hops. Dry the hops in the sun, and then reduce them to fine powder, by pounding in a mortar. Mix the soojee or flour and powdered hops with a little water, just sufficient to make a stiff dough; then add the sugar and knead all well together. Roll the leaven into a ball, wrap it lightly in a clean cloth, then in a blanket, and put it away for three days, when it will be ready for use.

N.B.—If worked up or kneaded once daily during the three days, the fermentation will be more perfect.

The above quantity will be sufficient for twenty-five pounds weight of bread.

## INDIAN PICKLES, CHUTNEES, SAUCES, ETC.

### Love-apple or Tomato Sauce

Ingredients:—Five hundred tomatoes; two pounds of green ginger, ground fine; a pound and a half of garlic, ground fine; one pound of chilies, ground fine; one pound of chilies, ground fine; one pound of salt; three pounds of tamarinds; and three quarts of vinegar.

Steep the tamarinds for twelve hours in a quart of the vinegar; strain them through a sieve, rejecting the stones, and add the other two quarts of vinegar, all the ground condiments, and salt; break the tomatoes into the mixture, and boil the whole, stirring it all the time until it thickens; remove it from the fire, and when cold strain it carefully and bottle the liquid, which is the sauce.

### Tomato or Love-apple Chutnee

Ingredients:—Two hundred large ripe love-apples, four ounces of raisins, seven ounces of salt, four ounces of sugar, eight ounces of chilies, finely sliced, four ounces of ground garlic, and seven ounces of ground mustard-seed.

Parboil the tomatoes in a quart of vinegar, add the other ingredients, and allow the whole to stand for ten to twelve hours; then boil it for twenty to thirty minutes over a slow fire; when cold, bottle it.

### Tapp Sauce

Ingredients:—Three seers or six pounds of peeled and sliced mangoes, two pounds of ground raisins, a pound of ground garlic, half a pound of ground chilies, a pound and a half of ground ginger, a pound of sugar, two pounds of salt, a quart of lime-juice, and six quarts of vinegar.

Mix all the above well together, put it into stone jars, and expose it to the sun for twenty days or a month, after which drain away the liquid, which is the sauce; boil it for ten to fifteen minutes, and when cold bottle and cork it.

### Sweet Chutnee

The refuse of the tapp sauce makes an excellent chutnee with the addition of some thick syrup, a few dried dates, a few more whole raisins, and some hot spices. Put the whole into a pan and let it simmer for a quarter of an hour, or until the syrup is absorbed and the chutnee reduced to a proper consistency; when cool, bottle, and cork it well down.

# Bibliography


Asiedu, J.J.: *Processing Tropical Crops,* MacMillan Press Ltd, London, 1989.

Axtell, B., Kocken E., and Sandhu, R., *Fruit Processing,* IT Publications, London, 1993.

Cammack, J.: *Basic Accounting for Small Groups,* Oxfam, Oxford, 1992.

Coultate, T.P.: *Food: the Chemistry of its Components,* Burlington House, London, 1984.

Fellows, P.: *Traditional Food Technologies,* IT Publications, London, 1997.

Hobbs, B. and Roberts, D.: *Food Poisoning and Food Hygiene,* Edward Arnold Ltd, London, 1987.

Howard, J.: *Safe Drinking Water,* Oxfam Technical Guide, Oxfam, 1979.

Ihekoronye, A.I., and Ngoddy, P.O., : *Integrated Food Science and Technology for the Tropics,* Macmillan Press Ltd., London, 1985.

Jagtiani, J., Chan, H.T. and Sakai, W.S.: *Tropical Fruit Processing,* Academic Press, San Diego, California, 1988.

Jay, J.M.: *Modern Food Microbiology,* D van Nostrand, New York, 1978.

M Harper,: *Small Business in the Third World,* J Wiley & Sons Ltd, Chichester, 1984.

Nelson, P.E. and Tressler, D.T.: *Fruit and Vegetable Juice Processing,* AVI Publications, Conn., 1982.

Rauch, G.H.: *Jam Manufacture,* Leonard Hills Books, London, 1965.

Sprenger, R.A.: *The Food Hygiene Handbook,* Highfield Publications, Doncaster, 1996.

Tindall, H.D.: *Vegetables in the Tropics,* MacMillan Press Ltd., London, 1983.

# Index

□□□